Advances in Religion, Cognitive Science, and Experimental Philosophy

Advances in Experimental Philosophy

Empirical and experimental philosophy is generating tremendous excitement, producing unexpected results that are challenging traditional philosophical methods. *Advances in Experimental Philosophy* responds to this trend, bringing together some of the most exciting voices in the field to understand the approach and measure its impact in contemporary philosophy. The result is a series that captures past and present developments and anticipates future research directions.

To provide in-depth examinations, each volume links experimental philosophy to a key philosophical area. They provide historical overviews alongside case studies, reviews of current problems and discussions of new directions. For upper-level undergraduates, postgraduates and professionals actively pursuing research in experimental philosophy these are essential resources.

Titles in the series include:

Advances in Experimental Epistemology, edited by James R. Beebe
Advances in Experimental Moral Psychology, edited by Hagop Sarkissian and Jennifer Cole Wright
Advances in Experimental Philosophy and Philosophical Methodology, edited by Jennifer Nado
Advances in Experimental Philosophy of Language, edited by Jussi Haukioja
Advances in Experimental Philosophy of Mind, edited by Justin Sytsma

Advances in Religion, Cognitive Science, and Experimental Philosophy

Edited by
Helen De Cruz and Ryan Nichols

Bloomsbury Academic
An imprint of Bloomsbury Publishing Plc
BLOOMSBURY
LONDON • OXFORD • NEW YORK • NEW DELHI • SYDNEY

Bloomsbury Academic
An imprint of Bloomsbury Publishing Plc

50 Bedford Square
London
WC1B 3DP
UK

1385 Broadway
New York
NY 10018
USA

www.bloomsbury.com

BLOOMSBURY and the Diana logo are trademarks of Bloomsbury Publishing Plc

First published 2016
Paperback edition first published 2017

British Library Cataloguing-in-Publication Data
A catalogue record for this book is available from the British Library.

ISBN: HB: 978-1-4742-2384-3
PB: 978-1-3500-4156-1
ePDF: 978-1-4742-2382-9
ePub: 978-1-4742-2383-6

Library of Congress Cataloging-in-Publication Data
Advances in religion, cognitive science, and experimental philosophy / edited by Helen De Cruz and Ryan Nichols.
pages cm. – (Advances in experimental philosophy)
Includes bibliographical references and index.
ISBN 978-1-4742-2384-3 – ISBN 978-1-4742-2383-6
1. Religion–Philosophy–Research.
2. Cognitive science. I. De Cruz, Helen, 1978- editor.
BL51.A424 2016
210–dc23
2015028705

Series: Advances in Experimental Philosophy

Typeset by Integra Software Services Pvt. Ltd.

Contents

List of Figures and Tables

Figures

Tables

1

Introduction: Cognitive Science of Religion and Its Philosophical Implications

Helen De Cruz and Ryan Nichols

Cognitive science and experimental philosophy

Philosophy of religion examines questions about the existence and nature of God and other religious entities and about religion as a cultural and cognitive phenomenon. Until the early modern period, philosophy of religion was one of the central areas of philosophical inquiry. Prominent philosophers of religion, such as Thomas Aquinas, believed that there were three sources on which we could rely in answer to questions about God and religious matters: our own reasoning skills, divine revelation (e.g., scriptures), and what other thinkers have said (tradition). These tools are still used in contemporary philosophy of religion, which places more emphasis on reasoning than on scriptures or tradition.

Commonly philosophers rely on reasoning from the armchair, that is, using their own intuitions in conceptual analysis, argumentation, and thought experiments. However, some philosophical branches, such as epistemology and philosophy of mind, have engaged increasingly with methods and results from the empirical sciences. Using a quantitative analysis, Knobe (2015) found that the majority of works by philosophers of mind from the 1960s to the 1990s relied on armchair reasoning, whereas of the papers published between 2009 and 2013, only a minority relied purely on armchair methods.

There are two ways in which philosophy can be empirically informed. It can be informed directly, by testing philosophical claims through experimental techniques (experimental philosophy). It can also be informed indirectly, by relying on the empirical findings of others (e.g., published studies in psychology, cosmology, or evolutionary biology) to support philosophical claims. Philosophical work that relies on scientific studies of other researchers is sometimes called "empirical philosophy" (Prinz 2008). This distinction between experimental and empirical philosophy might blur as experimental philosophy matures and comes closer to cognitive psychology in its methodology, but for now, it remains useful.

Experimental philosophy is a new and still controversial branch of philosophy, spanning diverse areas such as epistemology, moral philosophy, and metaphysics. Many experimental philosophers examine whether the intuitions of professional philosophers match those of nonphilosophers by employing experiments and surveys (Alexander et al. 2014). For example, Gettier cases (1963) provide evidence that knowledge is not merely justified true belief: when someone comes to believe it is 3 pm by reading her watch, and it is really 3 pm but that is a lucky coincidence as her watch has stopped, her belief is not knowledge. This is so even though her belief is both true and justified. Experimental philosophers (e.g., Weinberg et al. 2001) have found that this intuition is not universally shared among laypeople, and that there might be gender and ethnic differences in response to Gettier scenarios (Buckwalter and Stich 2014). However, in follow-up studies, this effect failed to replicate (see Kim and Yuan 2015, and Nagel et al. 2013 for a methodological critique and reply).

More recently, experimental philosophers have examined whether philosophers might enjoy expert knowledge not available to laypeople. For instance, Schwitzgebel and Cushman (2014) investigated whether professional ethicists might be less susceptible to ordering and framing effects in moral dilemmas compared to non-ethicists and laypeople. Philosophers were as susceptible as laypeople to such distorting effects. Consider trolley case vignettes. Participants are told to imagine a trolley speeding out of control, destined to kill several people who are tied to the track. Participants are commonly asked whether they would be willing instead to do one or the other of several options. They might be asked if they are willing to flip a lever to send the trolley down a different track that will kill only one person; willing to physically push with their hands a fat person over the railing and onto the track stopping the trolley but killing the fat person; or willing to push the fat person over the railing by using a pole to do so, etc. In the particular version of this experiment given to philosophers, they judged trolley scenarios more as equivalent if pushing a person off a bridge was presented first than if pulling a lever was presented first. In other words, philosophers were biased by ordering effects just like laypeople. The expertise hypothesis, which holds that philosophers would be less biased, is thus not empirically confirmed (see De Cruz 2015, Rini 2015, for analyses, and Schwitzgebel and Rust 2015 for a meta-analysis).

Empirical philosophy significantly predates experimental philosophy and can be traced back to early modern philosophers such as Hume, Reid, and Locke. Although these authors could not rely on controlled experimental results, they were nevertheless able to use less systematic empirical observations to test philosophical problems. For instance, Molyneux's problem (incorporated in Locke [1689] 1979) asks whether a person who was congenitally blind but could now see would be able to tell a cube from a sphere by sight alone. This was thought to have important implications for two major philosophical theories of the time: Rationalism and Empiricism. Natural philosophers in the eighteenth century believed that if a blind person newly given sight were to successfully visually discriminate between a cube and a sphere without touching those objects, then Rationalism must be true. (Their reasoning had to do with the unity of innate ideas across multiple sensory modalities.) Empiricists, however, typically predicted that such a person would not be able to discriminate between a cube and a sphere, or, if such a person could, it would only be under biased experimental conditions (see Nichols 2007, ch. 9, for this story). Philosophers like Diderot examined studies of patients with cataracts who were treated and visually presented with objects to draw conclusions about Molyneux' problem (see Degenaar 1996). Adam Smith theorized about conditions of disuse in the region of the brain processing visual information in patients who were given sight later in life (Smith [1751] 1982, 161). The empirical tradition of analysis of the Molyneux problem continues to this day, for example, recently by Held et al. (2011), who studied the capacities of newly sighted patients in India to recognize shapes by sight.

Empirical philosophy of religion dates to the eighteenth century, with authors like de Fontenelle and de Brosses, who, in the absence of careful observational fieldwork of anthropologists, used reports of religious beliefs, behaviors, and rituals from far-flung parts of the world. Today, the best known work from this period is Hume's *Natural History of Religion* ([1757] 1988), which unified available data into a theory about the psychological origins of theism, rooting it in anthropomorphism and early polytheistic traditions. Empirical philosophy is gaining ground with the increasing influence of naturalism and naturalistic methods, where philosophy is regarded as continuous with the sciences.

Two types of experimental input for the philosophy of religion: Cognitive science of religion and experimental philosophy of religion

Philosophy of religion can benefit from two types of experimental input, cognitive science of religion and experimental philosophy, leading to empirical philosophy of religion and experimental philosophy of religion, respectively.

The cognitive science of religion (CSR) is a relatively new, multidisciplinary field of the study of religious beliefs and practices. CSR scholars hold that religious beliefs and practices are typically the result of evolved, ordinary cognitive biases and processes. A common dictum in CSR is that religion is natural. Naturalness is a polysemic concept denoting that religious beliefs are easy to acquire and that they easily spread through cultural evolution (e.g., Boyer 2001), that beliefs that commonly occur in religious belief systems, such as belief in a life after death, occur early in development (e.g., Bering 2006), and that acquisition of religious beliefs does not require high cognitive effort, unlike say, scientific concepts (McCauley 2011). CSR has a broad scope, investigating topics such as religious rituals (e.g., Liénard and Boyer 2006), belief in spirit possession (Cohen 2007), and the psychosocial effects of belief in Hell (Shariff and Aknin 2014). Although CSR focuses on the cognitive correlates of religious belief, there is a large body of research on closely related topics as well, such as the relation between religiosity and prosociality (e.g., Norenzayan and Shariff 2008). For example, belief in hell but not heaven is positively correlated with low rates of social crime, whereas belief in heaven but not in hell is positively correlated with high rates of social crime (Shariff and Rhemtulla 2012). Although religion seems to make people more prosocial, this could be caused by non-religious factors, such as concern for one's reputation. Some experiments address such alternative explanations directly (Saroglou et al. 2005; Preston et al. 2010).

By comparison, there is very little work in experimental philosophy of religion (EPR), with only a handful of published studies to date (e.g., Nichols 2004; De Cruz in press; De Cruz and De Smedt this volume; Tobia in press). Compared to CSR, EPR is somewhat narrower in scope since it aims to have a direct relevance for philosophical questions. Many topics in CSR, such as spirit possession, fire walking, supernatural agency priming, and belief in reincarnation, are of only marginal interest (if at all) in philosophy of religion. Given the methodology of experimental philosophy, fewer methods are available to EPR compared to CSR. Although experimental philosophers have lately used methods such as eyetracking, most work is still conducted using written questionnaires. By contrast, CSR uses a wide variety of observational and controlled experimental techniques.

Empirically inclined philosophers of religion stand to benefit by familiarizing themselves with the methods and results of both EPR and CSR. But given the above discussion, these philosophers stand to improve their research and better understand answers to big questions about religion by looking first to CSR. First, the greater scope of CSR offers more resources to philosophers of religion. While the research goal of EPR is increasing our understanding of philosophical intuitions about religion, this is not an interdisciplinary pursuit and leaves philosophers mainly addressing familiar philosophical concerns rather than carving out new ideas. Second, methods in CSR tend to be stable and reliable in contrast to often simpler methods, experimental designs, and statistical testing found in experimental philosophy. Third, we see potential gains from applying insights from CSR to questions in philosophy of religion as of considerably greater importance than the insights that can be deduced from EPR results. Only methods and statistical analyses in CSR, not EPR, afford us the ability to understand influences, mechanisms, and robust correlations between findings about religious cognition, belief, and emotion.

It is advantageous for philosophers of religion to engage with CSR. Consider the philosophical question: Why do theists appear to deploy concepts of God that are inconsistent with their professed creedal beliefs? While philosophers debate whether or not a Christian monotheist is inconsistent when affirming two distinct concepts of God, Justin Barrett and Frank Keil investigated differences in the application of core theological concepts of God and God's properties from the perspective of cognitive psychology (Barrett and Keil 1996). In an "off-line" condition, participants reported their theological beliefs and did so in conformity with their stated beliefs. In an "on-line" condition, in which participants needed to remember details of a story, "adults who claimed that God can listen to or attend to any number of things at once, misremembered stories as saying that God was unable to hear something because of a loud noise or had to answer one prayer before going and attending to another" (Barrett 2007, 3). This was so despite the fact that these participants formally believe that God is not hard of hearing and is not restricted to one spatiotemporal location. In this case, CSR results shed light on the origins of an otherwise mysterious philosophical inconsistency among believers. This phenomenon was memorably dubbed "theological incorrectness" by Jason Slone (2004). Discussion has begun of the philosophical implications of this line of research for philosophy of religion and religious belief (De Cruz and De Smedt 2015, chapter 3).

Other features of CSR have received comparatively more attention by philosophers, perhaps because they are less narrowly religious issues. One feature of the debate about dualism in the philosophy of mind has been the subject of a number of papers in cognitive science, to wit the psychology of the so-called hard problem of consciousness. Philosophers have long appealed to intuitions and a priori reason to show that, "psycho-physical identity statements leave a significant explanatory gap" (Levine 1983, 355). But recently, philosophers, particularly those with one foot in cognitive science, have taken a psychological approach of this problem. By applying a dual-process theory of cognition to this explanatory gap, Fiala et al. (2012) have sought to elucidate why this gap appears unbridgeable to philosophers. They argue that an understanding of the cognitive psychology of self-directed thinking illuminates key differences found in thought experiments in which one is asked to think of oneself in the third person as opposed to the first person. Appreciating the importance of imagination and its limits—my inability to imagine that I myself do not exist as a conscious entity—sheds new light on a perplexing philosophical problem. This research assists philosophers of religion and of mind in understanding why, more generally, substance dualism has had a long appeal in the history of ideas.

The previous example shows that bringing together cognitive science and empirical philosophy can yield novel insights for philosophers. But a broad metaphilosophical issue arises just around the methodological corner: to what extent are science and religion compatible? Fueled by particularities of American history up through the present, debates about evolution and its incompatibility with religion can still be found in some parts of that country. Philosophers have focused on the metaphysical compatibility of scientific and religious claims, paying little attention to whether these claims are psychologically compatible. CSR takes a hands on approach. In a series of studies, Cristine Legare and colleagues found that people can easily accommodate and even integrate natural and supernatural explanations. For instance, Legare et al. (2012) examined how South African adults understood the causal factors involved in AIDS. Even though they understood the biological causes involved (viruses), they combined these scientific explanations with supernatural ones, for example, "witches can put one in the way of viruses." Although these studies do not directly address the metaphysical issue of the compatibility of science and religion, they indicate cognitive constraints on how people conceptualize the relationship between both fields.

We will now survey three domains in which philosophy of religion can benefit from empirical and experimental work in CSR: challenges to religious belief, the influence of cognitive constraints on religious beliefs and practices, and conceptualizing traditional issues in philosophical theology. We will show how the papers in this volume explore these issues.

Challenges from CSR to religious beliefs

A significant part of the philosophical literature that engages with CSR (e.g., Schloss and Murray 2009) is focused on the question of whether the results from CSR challenge religious beliefs: in the light of CSR results, are religious beliefs less likely to be reasonable, justified, or warranted? How, if at all, do CSR results challenge religious beliefs? There are three broad responses to the question of what, if anything, we can conclude from the evolutionary origins of religion and its psychological functions. *Evolutionary debunking arguments* propose that the psychological facts about religion, as uncovered by CSR, challenge the rationality of theism (e.g., Wilkins and Griffiths 2013). *Evolutionary justifications of religious belief* propose that CSR provides significant empirical support for theological claims (e.g., Clark and Barrett 2011). A third group of authors proposes that the etiology of religious beliefs is irrelevant for their epistemic standing (e.g., Thurow 2013). Three papers in this book (Wilkins, Teehan, Marsh and Marsh) propose that the evolutionary origins of religion challenge the rationality of religious beliefs.

John Wilkins explores similarities and differences between scientific and religious beliefs in the light of the evolutionary origins of human cognition. Both types of beliefs have biological, cultural, and sociological roots. Looking at these three etiological accounts, Wilkins argues that religion, but not science, is debunked by them. This is true regardless of whether religion is a by-product or an adaptation. If religion is adaptive because of its social utility, it does not matter whether it is true or false. By contrast, science can piggyback on the fact that common sense tracks truth to check results, and scientists go to great lengths to avoid error (replication, peer review process, etc.).

John Teehan looks in more detail at the problem of evil, a classic challenge to theism in philosophy of religion. This problem has received a lot of attention in mainstream philosophy of religion, and according to Teehan, cognitive science can further strengthen it. He shows this by considering the features of our evolved morality. Traditionally, the fact that humans have some unlearned (probably innate) sense of what is right and wrong has been regarded as evidence for theism; see, for example, formulations of the moral argument by Swinburne (2004). A crucial and often-overlooked feature of evolved morality is that it is an in-group adaptation. As a result, humans are more morally sensitive to those they consider in-group members, and conversely, they can be indifferent or cruel to those they consider part of the out-group. Teehan surveys empirical evidence for this, for instance, that people are less empathetic toward people who experience pain if they believe them to be out-group members (e.g., of different ethnic groups). This suggests that some moral evil, such as prejudice, between-group violence, and dehumanization, results from a properly working system of evolved moral cognition. This presents a challenge to theism. Teehan proceeds to consider some theodicies and argues that none is successful.

In the domain of religion, there is an enormous cross-cultural diversity of opinion, which has considerable skeptical force. For instance, Harris and Corriveau (2014) found that elementary school-age children are less confident about the existence of religious things, such as angels, than they are about the existence of endorsed scientific entities such as oxygen. They found that children's greater skepticism about religious entities is rooted in their awareness that people disagree about religious matters and have more consensus about scientific matters (with the exception of a few cases, such as climate change). In the philosophy of religion, this challenge from religious diversity has received some attention. Jason Marsh and Jon Marsh look at CSR for resources to develop a new challenge to religious belief from religious diversity, which they term the explanatory challenge. If God wants people to participate in the true theistic religion, why then would he permit multiple competing religious traditions to be successful? They show that religious diversity is inevitable under the main influential accounts of CSR: the way our minds work and the way humans aggregate in groups inevitably leads to religious diversity. In other words, since God is—under standard theistic accounts—the one who causes nature, and therefore, brains and environments like ours, divine design is responsible for religious diversity.

The influence of cognitive constraints on religious beliefs and practices

A central assumption of CSR is that religious beliefs and practices are the results of common, everyday cognitive processes that are widespread in humans. This emphasis on the mundane and on properly working cognitive functions marks an important difference with earlier psychological approaches to religion, which emphasized relatively rare features of religion, such as intense religious experiences and neurological conditions such as temporal lobe epilepsy (e.g., James 1902). It also marks a divergence with psychological accounts that regard religion almost exclusively as the result of unreliable cognitive processes, such as wishful thinking (e.g., Freud 1927; Hume [1757] 1988). CSR's emphasis on normal cognitive processes can be traced back to anthropological studies from the 1920s onward, when anthropologists started to gather detailed, first-hand observations of religious practices in other cultures. These observations shattered the view that religious practices were the result of a primitive mindset and suggested that there were common rationales underlying them. For instance, Malinowski ([1925] 1999) discovered that the Trobriand Islanders used rituals prior to deep-sea fishing, a highly risky mode of subsistence with uncertain outcomes, whereas they did not use rituals to help them in their everyday lagoon fishing, which was safe and had steady returns. This suggested that the use of magic was not irrational, but that there was some underlying rationale behind its use.

CSR authors combine anthropological observations with insights from contemporary cognitive psychology to explain rituals, why people perform them, and why they have particular forms. In spite of the wide diversity of rituals across cultures, there are cognitive constraints that govern their form, frequency, and distribution. For instance, Whitehouse (2004) found that religious rituals across cultures can be divided into two broad categories: infrequent, relatively uncodified high-arousal rituals that are stored in episodic memory, and more frequent, highly codified low-arousal rituals that are codified in semantic memory. In this volume, Cristine Legare, Rachel Watson-Jones, and Andre Souza examine the role of causal cognition in shaping religious rituals. Although rituals have long been conceptualized as ineffective by religious studies scholars, Legare et al. note that they are used, across cultures, to solve specific problems. They propose that rituals used for problem-solving purposes reflect intuitive beliefs about causation: people evaluate the efficacy of rituals using intuitive causal principles. Although rituals are causally opaque (i.e., there is no clear mechanistic explanation for why religious rituals are supposed to be causally effective), people can still rely on causal intuitions to gauge the purported efficacy of religious rituals. Using experimentally modified Brazilian *simpatias*, rituals that are used by nonexperts to bring about various goals, they found that rituals were perceived as more effective if they had a higher frequency of performance, a greater number of specific steps, and used religious icons. Rituals are often invoked to get an increased sense of control in situations that are beyond one's control. Legare et al. found that *simpatias* are regarded as more effective by Brazilian and US participants when they are primed with stimuli that invoke a sense of randomness.

Kelly James Clark explores the relationship between intuitive and reflective modes of thinking in atheism. Although most research in CSR has focused on the cognitive dispositions underlying religious belief, there is a small literature on the cognitive underpinnings of atheism. As Clark notes, some CSR research has the implicit normative assumption that atheism is rational, whereas theism is less than rational. This normative assumption is exemplified by research revealing a pair of correlations. The first finding is that religious belief is positively correlated with a series of incorrect responses to questions that activate intuitive thinking. The second finding is that the lack of religious belief positively correlates with correct responses to questions that activate reflective thinking. But, even if it is correct that there is a strong association between intuitive thinking and religious belief, and reflective thinking and atheism, does this indicate that theists are irrational, and atheists rational? Here, Clark first draws an empirical premise from the thrust of recent experimental philosophy to argue that the role of intuitions runs deep in philosophy of religion arguments. Clark then pairs this empirical claim with a follow-on conclusion, inspired by William James, that, in the case of philosophy of religion, we have little principled reason to choose atheist intuitions over theist intuitions.

Helen De Cruz and Johan De Smedt present results from a survey for philosophers, who were asked to rate eight arguments for theism (e.g., design argument, cosmological argument) and eight arguments against theism (e.g., argument from evil, argument from divine hiddenness). They found a strong correlation between how philosophers evaluated the arguments and their religious beliefs (theism, atheism, or agnosticism). Religious belief was in fact the strongest predictor of how participants evaluated the arguments. Having philosophy of religion as an area of specialization only had a modest effect on some of the arguments, for example, philosophers of religion evaluated the argument from divine hiddenness as stronger compared to philosophers with other specializations. Overall, there was no difference in how strong philosophers of religion found the arguments compared to nonspecialists. Philosophers of religion were just as strongly influenced by their religious beliefs as the other philosophers. De Cruz and De Smedt argue that their findings are consistent with a prominent role of confirmation bias in philosophy of religion.

Traditional issues in philosophical theology reconceptualized using empirical data from CSR

There are several conceptual questions in philosophical theology that deal with the nature of God and of human persons. What sort of being is God, and what attributes does he have? Are human persons embodied or disembodied? Philosophers of religion (e.g., Swinburne 1993, 2004) have examined how the so-called omni-properties of God, such as omniscience, omnipotence, omnipresence, and omnibenevolence, are related. Until recently, these issues were approached exclusively through conceptual analysis.

With the experiments of Barrett et al. (2001), CSR authors have begun to explore how people think about God and persons, and what light this can shed on philosophical concepts, such as the concept of omniscience. In these studies, Barrett showed that young children attribute omniscience to their parents and other agents before realizing that agents have cognitive limitations. For instance, when 3-year-olds are shown a cracker box that has rocks inside, they erroneously think their mother—who has not looked inside the box—will believe there are rocks in the box. Four-year-olds realize their mom will not know but will think there are crackers inside. This is a standard result of a false-belief task. Barrett et al. added a twist to the experiment, by also asking about God: children initially attributed omniscience to God and continue to do so at a later age. Barrett and Richert (2003) conclude that young children are "prepared" to think of God as omniscient; this belief is not the result of cultural learning but a cognitive default stance that they alter when they see other agents are really fallible in their beliefs. Because children continue to receive testimony that God is omniscient, they maintain that view. In a similar vein, younger children in a study by Giménez-Dasí et al. (2005) initially believed their best friend and God are immortal, whereas older children think their best friend is mortal but continue to attribute immortality to God. Thus, the theological concepts of divine eternity and impassiveness emerge at a young age. Debate continues about these results and about their apparent tension with earlier work that indicates that children also anthropomorphize God (see, e.g., De Cruz and De Smedt 2015, chapter 3). Yet, CSR is fruitful in identifying continuities between theological concepts and beliefs people intuitively hold.

Benjamin Purzycki and Rita McNamara explore how people across cultures conceptualize the minds of gods. Curiously, even in cultures where God is allegedly omniscient, people intuitively think that gods only care about very limited domains. These concerns are intimately tied to the problems generated by the social and natural environments in which people live. Purzycki and McNamara argue that the domains are closely related to behaviors that have fitness consequences, such as social behaviors and management of resources. Not all gods are concerned with what we would call morality (i.e., good and bad behavior). Morally concerned gods emerge in societies that are large-scale and that do not have effective secular instruments to deter potential free riders such as police. An ecological approach can explain why so many gods are concerned with etiquette, for example, wearing hats in the village. Since adherence to etiquette conveys social influence and control, it is understandable that gods (like their worshippers) would be concerned with violations of etiquette, as these may signal disrespect for social hierarchies and other important social institutions. Many gods are regarded as being concerned about rituals and their correct execution. Rituals are very public and reliable signals of commitment to the gods and social partners, which is why gods find them important, especially in times of conflicts over resources. Purzycki and McNamara also discuss nature, a domain many gods care about. Gods can be angered when people mismanage or pollute the environment. Religiously motivated resource management might contribute to fitness-enhancing behavior. Taken together, this research suggests that theological views about the concerns of gods are influenced by factors that have important consequences for human fitness.

Two papers in this volume focus on intuitive conceptualizations of personhood. Claire White, Robert Kelly, and Shaun Nichols present an empirical study among western participants who self-identify as spiritual, and who believe they have lived before their present life. This conviction is based on memories they have of past lives. This research ties into metaphysical questions about what a human person is, especially if one considers the possibility of multiple lives (reincarnation). What makes a present-day married woman with two children, working in an office, have the same identity as a male, childless shopkeeper in nineteenth-century Manhattan? According to Thomas Reid, personal (episodic) memories are key to personal identity: "Our own personal identity and continued existence, as far back as we remember anything distinctly … we know immediately, and not by reasoning. It seems, indeed, to be part of the testimony of memory. Everything we remember has such a relation to ourselves as to imply necessarily our existence at the time remembered" (Reid [1785] 1969, 586). White et al. show that episodic memories are indeed the most important factor to convince people they have had a past life, and moreover, that people feel a sense of personal ownership over the memory.

Whereas White et al. look at intuitive views of personhood in lives before the present one, Mitch Hodge considers afterlife beliefs, starting with the following philosophical puzzle: according to the Epicureans, it is not rational to fear one's death, since, when one is dead one ceases to exist, and therefore, there is no person anymore that can be harmed by the death. Some work in the psychology of religion is concerned with the fear of death and proposes that people are motivated to believe in an afterlife (either literally or symbolically) to assuage this terror. Hodge criticizes the undue emphasis on a person's *own* death, and the psychological literature's emphasis on the fear of one's own death. He argues that afterlife beliefs are spontaneously formed when we think about the deaths of *others*, not of ourselves. When we imagine what others who are not present are doing (e.g., a grandfather who lives far away), we make embodied representations of these people (e.g., he goes fishing). This form of social reasoning allows us to make inferences about what others are doing if they are not physically near us. According to Hodge, similar mechanisms operate when someone is deceased—we can still imagine grandfather fishing even though he is no longer alive; we spontaneously do so. In other words, we do not represent the death of others intuitively as annihilation but as a change in location. Even if our own death is nothing to us as the Epicureans maintained, the death of loved ones certainly is.

Directions for future research

The contributions to this volume explore different ways in which philosophy of religion can benefit from work in CSR. A vast domain of possible interactions remains unexplored. One of the most direct benefits of a more empirically informed philosophy of religion is a broadening of the field of study. As several critics have recently observed (e.g., Draper and Nichols 2013; Schilbrack 2014), philosophy of religion is at present narrowly focused on the topic of religious beliefs and tends to present generic or Christian monotheism and scientific naturalism as the only metaphysical options worthy of philosophical reflection.

To see how CSR could improve the practice of philosophy of religion, consider Schilbrack's (2014) recent manifesto. According to Schilbrack, philosophy of religion suffers from three interrelated problems: it is too intellectualist (focused on beliefs), too insular (not enough engagement with other disciplines), and too narrow in the range of religious traditions that are typically discussed. Note that even CSR needs to expand its frontiers, for example, to explain how emotion drives religious cognition. As it is, CSR avoids any discussion of the evolutionary and affective origins of cognition. This is not to say that it only focuses on beliefs. CSR authors have had a long-standing interest in religious rituals (Lawson and McCauley 1990; Whitehouse 2004; see also Legare et al. in this volume). CSR is an interdisciplinary endeavor and thus engages firmly with a variety of other disciplines. Its practitioners come from divergent scientific disciplines, including anthropology, developmental psychology, cognitive psychology, and philosophy. Finally, it has always had a broad interest in religious traditions outside of the monotheistic traditions that are typically discussed in philosophy of religion. Thus, by engaging with CSR, philosophers of religion stand to gain significantly in all these domains. We submit to you these essays with the conviction that reflection on them can catalyze positive change in both philosophy of religion and CSR.

Acknowledgments

We wish to thank James Beebe for inviting us to participate in this experimental philosophy book series. We are particularly grateful to the external anonymous reviewers who read drafts of the papers collected in this volume, and who provided valuable suggestions and feedback. We are also thankful to Johan De Smedt for comments to an earlier draft of this introduction. This project was financially supported by a postdoctoral research grant of the British Academy (grant pf130006) and by a subgrant from Templeton World Charities' "The Chinese Challenge" project.

Bibliography

Alexander, J., R. Mallon, and J. M. Weinberg. 2014. Accentuate the negative. In J. M. Knobe and S. Nichols, eds, *Experimental philosophy. Volume 2*, 31–50. Oxford: Oxford University Press.

Barrett, J. L. 2007. Cognitive science of religion: What is it and why is it? *Religion Compass* 1: 768–786.

Barrett, J. L. and F. C. Keil. 1996. Conceptualizing a nonnatural entity: Anthropomorphism in god concepts. *Cognitive Psychology* 31: 219–247.

Barrett, J. L. and R. A. Richert. 2003. Anthropomorphism or preparedness? Exploring children's God concepts. *Review of Religious Research* 44: 300–312.

Barrett, J. L., R. A. Richert, and A. Driesenga. 2001. God's beliefs versus mother's: The development of nonhuman agent concepts. *Child Development* 72: 50–65.

Bering, J. M. 2006. The folk psychology of souls. *Behavioral and Brain Sciences* 29: 453–462.

Boyer, P. 2001. *Religion explained: The evolutionary origins of religious thought.* New York: Basic Books.

Buckwalter, W. and S. Stich. 2014. Gender and philosophical intuition. In J. M. Knobe and S. Nichols, eds, *Experimental philosophy. Volume 2*, 307–346. Oxford: Oxford University Press.

Clark, K. J. and J. L. Barrett. 2011. Reidian religious epistemology and the cognitive science of religion. *Journal of the American Academy of Religion* 79: 639–675.

Cohen, E. 2007. *The mind possessed: The cognition of spirit possession in an Afro-Brazilian religious tradition*. Oxford: Oxford University Press.

De Cruz, H. 2015. Where philosophical intuitions come from. *Australasian Journal of Philosophy* 93: 233–249.

De Cruz, H. (in press). Religious disagreement: An empirical study among academic philosophers. *Episteme*.

De Cruz, H. and J. De Smedt. 2015. *A natural history of natural theology. The cognitive science of theology and philosophy of religion*. Cambridge, MA: MIT Press.

Degenaar, M. 1996. *Molyneux's problem: Three centuries of discussion on the perception of forms*. Dordrecht; Boston, MA: Kluwer Academic Publishers.

Draper, P. and R. Nichols. 2013. Diagnosing cognitive biases in philosophy of religion. *The Monist* 96: 420–444.

Fiala, B., A. Arico, and S. Nichols. 2012. On the psychological origins of dualism: Dual-process cognition and the explanatory gap. In E. G. Slingerland and M. Collard, eds, *Creating consilience: Integrating the sciences and the humanities*, 88–110. New York: Oxford University Press.

Freud, S. 1927. *Die Zukunft einer Illusion*. Leipzig: Internationaler Psychoanalytischer Verlag.

Gettier, E. L. 1963. Is justified true belief knowledge? *Analysis* 23: 121–123.

Giménez-Dasí, M., S. Guerrero, and P. L. Harris. 2005. Intimations of immortality and omniscience in early childhood. *European Journal of Developmental Psychology* 2: 285–297.

Harris, P. L. and K. H. Corriveau. 2014. Learning from testimony about religion and science. In E. J. Robinson and S. Einav, eds, *Trust and skepticism. Children's selective learning from testimony*, 28–41. Hove: Psychology Press.

Held, R., Y. Ostrovsky, B. Degelder, T. Gandhi, S. Ganesh, U. Mathur, and P. Sinha. 2011. The newly sighted fail to match seen with felt. *Nature Neuroscience* 14: 551–553.

Hume, D. [1757] 1998. *Natural history of religion*. In J. C. A. Gaskin, ed., *David Hume: Principal writings on religion*. Oxford: Oxford University Press.

James, W. 1902. *The varieties of religious experience: A study in human nature*. New York: Longmans, Green.

Kim, M. and Y. Yuan. 2015. No cross-cultural differences in the Gettier car case intuition: A replication study of Weinberg et al. 2001. *Episteme* 12: 355–361.

Knobe, J. 2015. Philosophers are doing something different now: Quantitative data. *Cognition* 135: 36–38.

Lawson, E. T. and R. McCauley. 1990. *Rethinking religion: Connecting culture and cognition*. Cambridge: Cambridge University Press.

Legare, C. H., E. M. Evans, K. S. Rosengren, and P. L. Harris. 2012. The coexistence of natural and supernatural explanations across cultures and development: Coexistence of natural and supernatural explanations. *Child Development* 83: 779–793.

Levine, J. 1983. Materialism and qualia: The explanatory gap. *Pacific Philosophical Quarterly* 64: 354–361.

Liénard, P. and P. Boyer. 2006. Whence collective rituals? A cultural selection model of ritualized behavior. *American Anthropologist* 108: 814–827.

Locke, J. [1689] 1979. *An essay concerning human understanding* (P. H. Nidditch, ed.). Oxford: Oxford University Press.

Malinowski, B. [1925] 1992. *Magic, science, and religion and other essays*. Prospect Heights, IL: Waveland Press.

McCauley, R. N. 2011. *Why religion is natural and science is not*. New York: Oxford University Press.

Nagel, J., V. San Juan, and R. A. Mar. 2013. Lay denial of knowledge for justified true beliefs. *Cognition* 123: 652–661.

Nichols, R. 2007. *Thomas Reid's theory of perception*. Oxford: Oxford University Press.

Nichols, S. 2004. Is religion what we want? Motivation and the cultural transmission of religious representations. *Journal of Cognition and Culture* 4: 347–371.

Norenzayan, A. and A. F. Shariff. 2008. The origin and evolution of religious prosociality. *Science* 322: 58–62.

Norenzayan, A., A. F. Shariff, W. M. Gervais, A. K. Willard, R. A. McNamara, E. Slingerland, and J. Henrich. in press. The cultural evolution of prosocial religions. *Behavioral and Brain Sciences*.

Preston, J. L., R. S. Ritter, and J. I. Hernandez. 2010. Principles of religious prosociality: A review and reformulation. *Social and Personality Psychology Compass* 4: 574–590.

Prinz, J. 2008. Empirical philosophy and philosophical intuition. In J. M. Knobe and S. Nichols, eds, *Experimental philosophy*, 189–208. Oxford: Oxford University Press.

Reid, T. [1785] 1969. *Essays on the intellectual powers of man*. Cambridge, MA: MIT Press.

Rini, R. 2015. How not to test for philosophical expertise. *Synthese* 192: 431–452.

Saroglou, V., I. Pichon, L. Trompette, M. Verschueren, and R. Dernelle. 2005. Prosocial behavior and religion: New evidence based on projective measures and peer ratings. *Journal for the Scientific Study of Religion* 44: 323–348.

Schilbrack, K. 2014. *Philosophy and the study of religions: A manifesto*. Malden, MA: Wiley-Blackwell.

Schloss, J. P. and M. Murray, eds. 2009. *The believing primate: Scientific, philosophical, and theological reflections on the origin of religion*. Oxford: Oxford University Press.

Schwitzgebel, E. and F. Cushman. 2014. Expertise in moral reasoning? Order effects on moral judgment in professional philosophers and non-philosophers. In J. M. Knobe and S. Nichols, eds, *Experimental philosophy. Volume 2*, 347–366. Oxford: Oxford University Press.

Schwitzgebel, E. and J. Rust. in press. The behavior of ethicists. In J. Systma and W. Buckwalter, eds, *The Blackwell companion to experimental philosophy*. Malden, MA: Wiley-Blackwell.

Shariff, A. F. and L. B. Aknin. 2014. The emotional toll of Hell: Cross-national and experimental evidence for the negative well-being effects of Hell beliefs. *PLoS ONE* 9: e85251.

Shariff, A. F. and M. Rhemtulla. 2012. Divergent effects of beliefs in Heaven and Hell on national crime rates. *PLoS ONE* 7: e39048.

Slone, D. J. 2004. *Theological incorrectness: Why religious people believe what they shouldn't.* Oxford: Oxford University Press.

Smith, A. [1751] 1982. *Essays on philosophical subjects.* Indianapolis: Liberty Classics.

Swinburne, R. 1993. *The coherence of theism* (rev. ed.). Oxford: Clarendon Press.

Swinburne, R. 2004. *The existence of God* (2nd ed.). Oxford: Clarendon Press.

Thurow, J. C. 2013. Does cognitive science show belief in God to be irrational? The epistemic consequences of the cognitive science of religion. *International Journal for Philosophy of Religion* 74: 77–98.

Tobia, K. in press. Does religious belief infect philosophical analysis? *Religion, Brain and Behavior.*

Weinberg, J. M., S. Nichols, and S. Stich. 2001. Normativity and epistemic intuitions. *Philosophical Topics* 29: 429–460.

Whitehouse, H. 2004. *Modes of religiosity: A cognitive theory of religious transmission.* Walnut Creek, CA: AltaMira Press.

Wilkins, J. S. and P. E. Griffiths. 2013. Evolutionary debunking arguments in three domains: Fact, value, and religion. In G. W. Dawes and J. Maclaurin, eds, *A new science of religion*, 133–146. New York: Routledge.

2

Is Religion or Science Debunked by the Evolution of Cognitive Faculties?

John S. Wilkins

Introduction

In this chapter, I wish to compare and contrast the role of naturalizing explanations of both science and religion, in order to show how evolutionary debunking arguments apply differently to science and religion. This will shed light, I hope, upon the role that environmental selection pressures (either at the genetic level or the cultural level; what applies at one level will mutatis mutandis. apply at all others) play in such beliefs tracking the truth of the content of beliefs. I will conclude that religion is debunked to a degree by naturalizing explanations in terms of evolution, while science, generally, is not.

Evolutionary debunking arguments

Evolutionary debunking arguments (EDAs) have been widely applied to moral beliefs, either by arguing that moral realism (the claim that moral claims are facts) is an "error," for instance, the error theory of John Mackie, (Mackie 1990), or defending moral realism against skepticism (Kahane 2011). Recently, EDAs have been applied to religion and knowledge claims generally (Ruiz 2013).

An evolutionary debunking argument has the following form (Kahane 2011):

Causal premise. *S*'s belief that *p* is explained by *X*
Epistemic premise. *X* is an off-track process
Therefore
S's belief that *p* is unjustified.

An "off-track process" is a process of the evolution by natural selection of cognitive dispositions or belief acquisition, in which what makes the dispositions or beliefs (for simplicity we shall refer to beliefs) *fit* is decoupled from its *truth*. For example, suppose that belief in God is a fitness-enhancer in a given social environment, because it increases one's chances of reciprocal altruism in times of need in societies where kin relationships are unknown or too attenuated (Wilkins 2015). The fitness here is granted in virtue of the mere fact of *having* the same beliefs as those of your social milieu, not in virtue of the truth of those beliefs. In fact, even if the beliefs are entirely false, fitness is still conferred to the believer. Griffiths and I (Wilkins and Griffiths 2013) have proposed that this sets up a "Milvian Bridge" problem. The Milvian Bridge was the site of the decisive battle between Constantine, with his largely Christian troops, and Maxentius, with his largely "pagan" troops. Later Christian writers took the success of this battle to indicate the truth of the Christian religion, when in fact it may have been that the social cohesion of the beliefs themselves acted to motivate the fighters—hence the fitness (success in battle) is decoupled from the truth of the beliefs. Had Constantine's troops been largely Mithraists, however, the social cohesion of *their* beliefs may very well have resulted in a victory also. Hence, the truth of the beliefs is an "off-track" process.

This raises an issue for the beliefs of science.[1] *Mutatis mutandis*, what is true of religious and moral beliefs might also be true of scientific beliefs, which may confer fitness upon their bearers for reasons of social benefit rather than their truth. Hence, as is sometimes made out, a scientific belief might be fitness-conferring because it suits the socioeconomic interests of the specialists (by improving their professional career, or serving the interests of those capitalists or imperialists who fund science). In order to defend the truth-like nature of our best scientific theories, we proposed that scientific beliefs, and indeed environmental beliefs in general, do confer fitness because they are (approximately) true. Science, we argued, is an "on-track" process. Religion, we aver, is not.

The Milvian Bridge: Its construction and utility

We define a Milvian Bridge as:

> The *X* facts are related to the evolutionary success of *X* beliefs in such a way that it is reasonable to accept and act on *X* beliefs produced by our evolved cognitive faculties.[2]

A bridge is an instrumental object. It has to support the traffic placed upon it without collapsing or failing in extreme environmental conditions (recall the famous Tacoma Narrows bridge that, when hit by extreme winds at the right speed, oscillated into catastrophic failure in 1940 due to the interference matching its periodic frequency). Hence, the principles used by engineers to build a bridge must in some respect match the true physical conditions of the bridge and its environment. The engineers' beliefs must track the physical properties employed in the bridge's design and construction and the physical conditions it will encounter. How do engineers acquire these beliefs? We can assume that no engineer was born knowing these facts, and so, the education they acquire in school and college must itself teach these true facts. Nor are they commonsense beliefs, or else, the Tacoma Narrows bridge would not have been built the way it was. Engineering, of course, is the outcome of a process of social evolution.[3] What works is copied and what fails is revised, and an elaborate system of principles and both overt and tacit knowledge has been built up over generations, so that a graduating civil engineer does not repeat the mistakes of the past. Of course, they may very well make novel mistakes, but solutions to them will also be folded into the body of knowledge, and so on.

What is true of engineering principles is also true of scientific principles. Solutions are proposed and tried out, and those that survive are retained to become foundational knowledge for subsequent investigations. Of course, this is not an inevitable or monotonic process; there are false avenues, beliefs retained for contingent cultural or historical reasons that have never been tested, but overall, science has progressed since it began. To adapt a saying of Hacking's (Hacking 1983), if we can manipulate the theoretical objects of science, they are real. We have the capacity now to manipulate subatomic particles, genetic variants, and the very molecular structure of materials. That ain't hay. But there is a long distance between scientific theories and principles and our evolutionary inheritance of cognitive faculties. So we began by arguing that our commonsense world, our *Umwelt*, as von Uexkull (1926) called it, is itself an on-track outcome. According to this account, the recognition of middle-sized objects, including other organisms, potential food, and geographical features like cliffs and trees, is the result of selection for the ability to track our environment at a scale that increases the fitness of the organism that can do this. We distinguish between social and environmental cognition, to attend to the reason why some cognition is truth-tracking.

Consider an asocial organism that is capable of forming beliefs. Every belief it acquires based upon its inherited dispositions to respond to environmental stimuli and form doxastic stances to these stimuli will be fitness-relevant. Some beliefs will be fitness-lowering (believing there is no predator when there is one), some fitness-enhancing (believing that there is a cliff when there is one), and so on. Not all of its beliefs will be true, and there will be an error rate of false positives (e.g., hyper-vigilance for predators) and false negatives (e.g., failures to spot the cliff in time to avoid falling over it). However, a population of these organisms will tend to trade off errors to get the best outcome of true beliefs that enhances the overall fitness of the group. Overall, then, the *Umwelt* of these organisms will tend to track truth, by which we mean form beliefs that are instrumentally correct in their lives.

Now, if we consider a social organism, their environment also includes other members of the population, and what they react to. What those other organisms believe affects the degree of social interaction and the nature of those interactions. As an example, a bird whose species has a learned song but which has learned a song that differs substantially from other members of its population will be unable to mate and perhaps be excluded from foraging within the population's range. So this is a kind of *social tracking*, and there is no truth that is tracked other than the fact of the song (the beliefs and dispositions) of the other individuals in the population. Social stimuli beliefs do not track anything other than themselves. What counts for fitness is that you sing the right song, not that the song represents the world apart from the social milieu correctly. If religion or morality is fitness-enhancing, this is how they confer it. But no similar story can be made out for environmental fitness; beliefs that are wrong get their bearers dead or injured. Commonsense reasoning is therefore more or less truth-tracking. However, many beliefs formed by common sense are false. Not *all* of them can be, of course, or the survival of the agent that acts upon them would be a miracle, and miracle arguments are unsatisfying. However, common sense is no guarantee of true beliefs. The beliefs it generates either need to be tested for false positives or negatives, or else, they make no factual claims as such. Science is founded upon common sense, although it is not generally true, as T. H. Huxley said, that science is *just* organized common sense. Primates evolved to deal with aspects of their environment that do not include quarks, quantum mechanics and long-term populational changes and ensembles. Nevertheless, primate common sense is sufficient to start with. We can see and roughly measure all kinds of mesoscale objects and recall how they behaved in the past.

An organism with *environmental* beliefs must not acquire its success accidentally, nor may scientific hypotheses. While it is true that correlation does not imply causation, it must at the very least indicate strongly that something causal is in play, and the more elaborate the beliefs and the more widely reliable they are, the greater the warrant that the beliefs are, if not true in a philosophically satisfying sense, true in an instrumental sense of greater precision and accuracy. We have *prima facie* warrant that complex hypotheses that apply in a large range of cases for which the hypothesis is intended to apply are true to some degree. Our asocial organism has true beliefs just to the extent that those beliefs undergird success in the environment in which they are acquired. Beliefs about cliffs do not have fitness-enhancing effects in flat environments, but they do in rugged landscapes.

Counterfactual arguments in which success accidentally correlates to beliefs, such as Plantinga's "cuddly tiger" response, where the best way to pet a tiger is to run away from it, coincidentally preserving the person who has these strange beliefs about tigers,[4] leave open the possibility of false beliefs about the environment increasing fitness. However, they do not have sufficient purchase when an entire population is involved. It is correct to say that many organisms have no intensional beliefs that motivate their fitness-enhancing behaviors (which may be hard-wired neural responses to certain stimuli such as sudden movement that confer the fitness), but in those that *do* have motivating beliefs, the likelihood that all beliefs are factually false and yet enhance fitness beggars belief. *Some* beliefs may be false, but not *all* can be.

This scope insensitivity is at the heart of Plantinga's *Evolutionary Argument Against Naturalism*, and it cannot be maintained. He argues from the conceptual possibility of a single belief being false but fit, like the cuddly tiger belief, to the view that all commonsense and scientific beliefs may be (of a particular kind, as in scientific beliefs about evolution itself). Unless there is some very arcane accidental combination of false beliefs that may generally increase fitness, it is likely that a belief set must on the whole be true for environmental beliefs. The greater the complexity of the successful beliefs, the less likely it is that an accident "explains" their success; in fact, it would be another miracle if a system of mostly or all false beliefs was successful.[5] The presumption must be, then, that if environmental beliefs increase fitness, it is most likely that they do so, once we bracket out social fitness, by providing reliable beliefs that are in some sense true. Parenthetically, it is ironic and telling that to criticize evolutionary naturalism, Plantinga needs to posit a miraculous scenario.

Some might worry that our approach fails because it will not convince the radical skeptic posited by Plantinga, and there is a sense in which this is quite correct. Some skepticism will never be resolved from the environmental utility of beliefs. Radical skeptics allow thought experiments in which all beliefs are false but useful. Plantinga, for instance, wants to argue that one might have all false beliefs which coincidentally lead to survival, and hence one cannot ground naturalism upon the success of (scientific) beliefs. His argument is a form of *tu quoque*: if evolution is true, then one ought not to be a naturalist about evolved beliefs, because naturalism cannot be supported merely by the success of beliefs. But we are not attempting to counter radical skepticism. Instead, our worry is about whether science and the commonsense beliefs that science rests upon are subject to an EDA. Consider a parallel case: vision. One might argue that since vision is subject to illusions, one must not expect vision to give us a picture, as it were, of what is before us. That would be the radical skeptic's argument. But the analogous argument we might offer to our own is whether vision can be relied upon in ordinary cases, given that it has some error in cases our visual system did not evolve to deal with, like the Lyer-Müller illusion. This deals with straight lines and angles that are rarely, if ever, found in nature, and so, our failure in such cases is not a debunking of the ordinary reliability of our vision. It sets the limit cases of the reliability of vision. It does not require nor imply that all vision is misleading.

That there are (some) false beliefs in common sense or science is not a rebuttal of the claim that (most) beliefs in common sense or science are true in virtue of their having been selected for truth-tracking. One would need to show that all or most beliefs are false to reduce commonsense and scientific beliefs to the same status as the debunked moral and religious realisms. Thought experiments (designed by radical skeptics) will not do. Even if we adopt something like the structural realist view (Psillos 1999), or at the opposite end, the antirealism of van Fraassen (Van Fraassen 1980) and others, where the content is not necessarily true, so long as the theories so generated track the *structure* of the world,[6] they are truth-tracking nevertheless. This even holds for implicit theories (the monkey that reaches for the ripe fruit may have no explicit beliefs about fruit; or a societal norm may track real world facts with beliefs about, say, the reasons why crops grow, that are false or nonexistent). If the intensional content has implications that lead one to behave in ways that do not track truth, however (like exorcising those with a fever), this fails.

Warrant and science

One might think we leap too quickly from the truth-tracking of commonsense beliefs to the truth-tracking of scientific beliefs. Griffiths and I did not offer a detailed extension of our argument from common sense to science, but this does not mean that the Milvian bridge cannot be extended to science. We argued that the reliability of commonsense beliefs overall allows us to debug errors in our higher-level scientific beliefs. This is widely accepted. For example, Ian Hacking argued that the reliability of the telescope and the microscope could be "fact checked" against naked eye observations (of far objects up close, and of small objects by keen vision), thus giving warrant to observations that could not be made by the naked eye (Hacking 1983). By a process of cross-checking observations using a variety of techniques (such as testing electron tunneling microscopes against targets that can be investigated through other means, such as X-ray crystallography for proteins, and light microscopy for cells, etc.), and by testing an assay against many different samples of different kinds, we can evaluate the efficacy of observational tools, by reducing the likelihood that the reliability of these observations is due to chance, since consistent efficacy would be another miracle argument.

Our commonsense observations also underpin a lot of ordinary science. For example, the investigation into field biology and ecology relies on tools that are themselves not terribly theoretical or exotic, like binoculars, buckets, and spades. If commonsense observation is reliable in the main, we can learn both how more distal observational techniques work (and more to the point, when they do not) and how our commonsense observations sometimes fail. Yes, we sometimes fall prey to illusions like Lyer-Müller, but that can be checked by the simple use of a piece of string or ruler. We can, in short, use common sense to test common sense as well as science. Finally, one obvious way in which commonsense cognition can be used to test science is to see, plainly, where science makes predictions that are false. Much scientific prediction is highly theoretical and assay-relative, to be sure, but some is simply a matter of looking to see if a result the theory says should occur does so. We are not implying a naive theory-observation language distinction, although this is not so easily dismissed as the post-logical positivist tradition of the philosophy of science seems to think (Quine 1993); but we merely assert that commonsense observations can have an appreciable role in establishing the warrant for parts of scientific theories.

Another difference with the nature of science is that it strives, at great cost, to eliminate type I (false positive) and type II (false negative) errors. There is a cost to this. As a result, good science is costly. Not only do we need to have hypothesis generation, we also need to have testing, retesting, and time-costly analysis. This gives rise to a Big Data Problem (Frankel and Reid 2008)—as more information is acquired, our ability to clean the data, interpret the meaningful data from the noise and measurement error, and to remove signals from older, less reliable or precise, data is correspondingly reduced. The result is that science is hard, more reliable, and relatively decoupled from the errors of reasoning that commonsense reasoning is heir to.

Although the practice of science through such things as peer review and replication studies is relatively recent, and relatively flawed as a way of eliminating error in absolute terms, the fact that studies are peer reviewed and replicated indicates a stronger tendency to eliminate error than commonsense epistemology. In particular, the use of, and debates over, statistical methodology has reduced much of the error in scientific results, although bad statistics leads to undue confidence in hypotheses, particularly in the biosocial sciences (Cumming et al. 2007). In paying the cost of error elimination up front, as it were, science is not only truth-tracking, it is truth packing. But, this is not some difference in kind: it is merely the extreme tail of the distribution of error costs in signaling (Godfrey-Smith 1991).

Genealogical explanations of religion

Having thus framed science as a relatively undebunked evolved belief-generating process, let us now contrast that with religion, or at least some aspects of religion. First, this leads us to ask what counts as "religion"? Pascal Boyer (Boyer 2001) lists four kinds of explanations of religion that may help us classify what counts as a religion such that we might need to explain it naturalistically: Religion provides explanations (I will call this the *etiological* account); religion provides comfort (the *existential* account); religion provides social order (the *sociological* account); and religion is a cognitive illusion (I would prefer to say, a cognitive by-product; the *cognitive* account).[7] We are concerned here with the cognitive aspect of religion, as a belief-producing process. However, at least one sociological account has been given of the belief-producing aspect of religion: Marx's notion of "false consciousness" and its heirs and successors. This is also a form of naturalizing of religion and is also a debunking account (religion is the opiate, that is to say, the analgesic, of the masses). But for an EDA or close analogue to take effect, we need to focus upon historical, or indeed *genealogical*, accounts of religion: accounts which explain the nature of religion in terms of its singular origins.[8]

Evolutionary accounts are only one kind of genealogical account, and evolutionary accounts include both sociological and biological explanations of religion. One thing they all have in common, however, is the assumption of functionalism: is religion adaptive or not? (sociologically or biologically) is the key question. There are three kinds of emphasis in such questions (Table 2.1). If the question is "is religion adaptive?," we might choose to ask is it *adaptive* or not, since maybe it is a side-effect or spandrel of adaptive traits, or maybe it is simply the result of inherited phylogenetic traits which in humans are not subjected to selective pressures—I make a limited case for this elsewhere (Wilkins 2015). Or we might put the emphasis on *religion* and ask whether in fact this term denotes a single phenomenon or process. Religion is notoriously hard to define and has been for 200 years, since scholars began to take non-Abrahamic religious traditions in their own terms. If there is no single thing, then there can be no single account in terms of selective pressures. Or we might ask whether or not what religion is selectively advantageous *to* is indeed a single set of adaptive landscapes and pressures, or a response to variable ecologies. In short, what is religion's *ecology*? Is it civilization, empire, trade, social class, and so on? We might also ask in what *way* religion adapts: is it group selection, kin selection, individual selection, or gene/meme selection? All the major issues of the debate over levels of selection in biology apply also in the case of religion.

Now, we can bracket out most of these, because if religion is adaptive, it really does not matter much to the question of the truth-tracking of religious beliefs whether those beliefs are the result of group cognition, individual cognition, or (sub-brain) modular cognition. But the question of whether religious beliefs are of a single kind, and if so, what they adapt to, is crucial. I am going to have to make some simplifying assumptions. I shall assume that we speak of a single type of belief: nonempirically supported metaphysical beliefs.

Of course, there are religious beliefs that have empirical foundations in whole or in part (particularly the folk psychological assumptions about love, fear, loyalty, etc.). By "empirical foundation," I do not mean empirical support, however; only that these beliefs may act in the same fashion as commonsense realism, and either lead to more refined and tested beliefs or have some general

Table 2.1 Issues in the question "Is religion adaptive?"

Religion	is/is not adaptive	is/is not a single phenomenon
To an ecology:	Specify natural or social selective factors	Specify kind of phenomena denoted
At level:	At a group level	Group shared beliefs/institutions
	At a lower individual level	Individual experiences/behaviors

truth-tracking function. These can be subsumed under common sense to the extent that they have such functions, so let us leave them out of consideration.

We might also ask in what way religious beliefs are adaptive. There are several active hypotheses for religious beliefs (Bulbulia 2004): they are side effects of our evolved ability to "mind-read" the intentions of others (agency detection working overtime: Barrett and Malley 2007); they are a kind of cognitive superstimulus based upon attributions of counterintuitive powers to agents; they are useful for norm enforcement in societies (Shariff and Norenzayan 2007); and of course, they can be both biologically nonadaptive or maladaptive while being socially or culturally adaptive. But of the adaptive explanations, whether biologically based or socioculturally based, none of these explanations rely upon the truth of the beliefs as such but rely upon their utility to achieve some biocultural end, like group cohesion or economic outcomes. Even the agency detection account made popular by Barrett is explicitly a false positive-delivering process (Barrett 2013)—that is to say, it is an error of an adaptive system. And we can treat implicit beliefs (such as the rightness of modes of talking about gods in particular ways, or the necessity for certain rituals) as potentially explicit beliefs about the world and the gods (if you do not say "Bless you!" after a sneeze, the devil can enter the sneezer; etc.).

We can dismiss the truth-tracking of spandrel-derived beliefs immediately (with a caveat below). If they are spandrels, and have not been modified by a process of evolution by selection, natural or cultural, then they are not adaptive and hence do not track truth. The caveat is that religious beliefs, or more properly religious rituals, may encode a degree of ecological information (say, about when to plant or irrigate crops) that is independent of the truth of the justification or explications—that is, the content—of these rituals (e.g., because it is Shiva's birthday), and in this case, an EDA is unnecessary to debunk the beliefs, even if the rituals have social utility, since it is the structure and rehearsal of the rituals that confers the fitness, not the content of the beliefs associated with them. However, in the cases of the adaptive nature of religious beliefs, it is never the case that in a naturalistic account what the beliefs have adapted to is the intension, that is the content, of the belief: the Christian nations of the world are not Christian because interactions with God made non-Christian nations less fit, but because having non-Christian beliefs made the non-Christians less fit in the face of political, economic, cultural, and above all military superior Christian force (*m.m* for other religions).

Some Christian theology makes much of Paul's comment that all people know God (Romans 1: 18–20), the basis for the doctrine of the *sensus divinitatis* in Calvin's *Institutes*. As Calvin put it, "since from the beginning of the world there has been no region, no city, in short, no household, that could do without religion, there lies in this a tacit confession of a sense of deity inscribed in the hearts of all" (I.3.1). Various theologically inclined philosophers have made of this the claim that there are no atheists, since in order to reason at all they must have this sense of god, and thus in order to be atheists they must suppress that sense. That this is question begging is obvious, but let us ask in what way having that sense could itself be the outcome of selection. How could the *sensus divinitatis* evolve? A naturalistic account would have it that this is an effect of some other adaptive cognitive feature (my preferred explanation is that it is a side effect of social dominance psychology and the need to feel one has a place in one's troop, which is part of our evolved mating strategy and social organization), but in order for the content to be true, there would need to have been empirical consequences to a lack of that sense such that those who do not have it, or have an impaired version, are less fit. One way might be that the gods themselves smite those who fail to revere them; this, after all, was the view of classical Hellenic mythologies among many others. Death by deity would be a great fitness-lowerer. However, given that the deity not revered changes in every instance, across cultures and eras, one must posit either a deity who simply hates atheists no matter what deity various believers adopt, or that all the deities are real (henotheism) and punish "their" peoples. Neither solution will serve the modern monotheistic faiths, although Vedic religion may be able to accommodate both.

Darwin's horrid doubt

What kinds of claims are not truth-tracking in general? Darwin himself may have given the lead on this, and how it has been dealt with is instructive. Darwin wrote to William Graham of the view that religious reasoning was reliable:

> ...I have had no practice in abstract reasoning, and I may be all astray. Nevertheless you have expressed my inward conviction, though far more vividly and clearly than I could have done, that the Universe is not the result of chance. But then with me the horrid doubt always arises whether the convictions of man's mind, which has been developed from the mind of the lower animals, are of any value or at all trustworthy. Would any one trust in the convictions of a monkey's mind, if there are any convictions in such a mind? [July 3, 1881, in (Darwin 1888)]

Call this the *modified monkey brain objection*. It is worth noting, though, that Darwin did not make the same argument for the conclusions of science immediately before this passage, and after it. Evolution gives no comfort to those wishing to find a naturalistic foundation for its claims.

Plantinga (2008) makes much of this comment of Darwin's. But what neither he nor his commentator in the same journal (Mohrhoff 2008) go on to note is what *subject* Darwin is worrying over. He prefaced the quoted comments with this:

> ... there are some points in your book [Graham's *Creed of Science*] which I cannot digest. The chief one is that the existence of so-called natural laws implies purpose. I cannot see this. Not to mention that many expect that the several great laws will some day be found to follow inevitably from some one single law, yet taking the laws as we now know them, and look at the moon, where the law of gravitation—and no doubt of the conservation of energy—of the atomic theory, &c. &c., hold good, and I cannot see that there is then necessarily any purpose. Would there be purpose if the lowest organisms alone, destitute of consciousness existed in the moon? But I have had no practice in abstract reasoning, and I may be all astray. Nevertheless you have expressed my inward conviction, though far more vividly and clearly than I could have done, that the Universe is not the result of chance.

Only *then* does he suggest that a modified monkey's brain is inadequate to the task of reasoning, about *divinity and design*. The irony is that Darwin in no way worried that such a brain was adequate to the task of uncovering the laws of nature themselves. This was rather disingenuous of Plantinga and points up both Darwin's rather careful abstract reasoning despite his disclaimer and the fact that modified monkey brains may learn quite a lot about the world in an evolved naturalistic manner. Why is this?

Darwin is making the point that what monkey brains, and by extension our brains, evolved to deal with by natural selection was the environment for the bodies that carried those brains. Metaphysical claims therefore cannot have contributed to the evolutionary success of the bearers of those brains. Since religious beliefs as I have defined them consist solely of such metaphysical claims, neither can they. In fact, metaphysical claims in general cannot have contributed to ecological fitness; however, they may contribute to social fitness. One might argue that natural reason gives a more reliable account if we are dealing with philosophical versions of religion, such as the famous Spinozan deity which Darwin seems to have accepted, along with such luminaries as Einstein and Schleiermacher (Mason 1997). However, here, too, the problem arises that there is no reason to think that human reason evolved response to the success of inferences about supernatural beings or metaphysical issues. The elite forms of religion that are represented by theologians like Tillich, Bultmann, or Cupitt are no more immune from the EDA than the folk forms of religion that Dawkins attacks (Dawkins 2006). One's cognitive fitness does not depend here upon the truth of the content of these beliefs. Parenthetically, this also means EDA arguments apply to intuitive metaphysics in general, not merely religious ones.

Truth-tracking in human evolution

Human cognitive evolution evolved to deal with problems of enhancing fitness in several ways. One is *social tracking*. This is the ability to track social mores, in order to be accepted as a part of the community in order to benefit from reciprocal support or reciprocal altruism (Boyd and Richerson 1988). As I have argued, in social contexts, the content of these mores is irrelevant, so long as the mere *having* of the beliefs is sufficient to motivate others to aid you in tough times and accept you in the ordinary times. Another is *psychological tracking*. Here, the fitness-enhancing is done by correctly anticipating what other people mean or intend and to evade deception. This is sometimes called "intention-reading," or "theory of mind" (Leslie 1987). Here, the fitness is imputed by *correct* beliefs about the intentions of others; other people's intentions therefore play a role as selective events in the environment of the person doing the intention-reading. But the contextual elements of intentional ascriptions depend very much upon cultural constructs. Someone seeing the Maori *Haka* in a rugby match interprets the New Zealand players' intentions very differently from how members of an opposing force would have interpreted the performers a century or more ago, as it was then a demonstration of the willingness to engage in deadly force.

At the other extreme, engaging in interactions with the environment, dealing with predators of course, but also being able to interpret the affordances of the ecosystem and objects within it, involves little more than truth-tracking—although the intentions of the "engineers" who pick up a piece of wood and sharpen it to make a spear may be quite constructed. So we have cases where selection occurs because of a failure to show group commitment, where it occurs because of a failure to interpret others' intentions correctly, or where it occurs because of a failure to read the environment correctly. Each case tracks a truth, but only the environmental truth matches up with the belief content of the cognitive agent.

It has been argued recently that even reasoning itself did not evolve in order to reach reasoned conclusions but instead to convince others by argumentation and so gain standing in one's community (Mercier and Sperber 2011):

> For communication to have evolved, it had to be advantageous to both communicators and receivers (who are, of course, the same individuals but acting in two different capacities). What makes communication advantageous to receivers is that it provides them with rich information that they could not, or not easily, have obtained on their own. For this, the information they receive has to be genuine information; that is, *close enough to truth*. What makes communication advantageous to communicators is that it allows them to achieve some desirable effect in the receivers. For this, the information they emit has to be conducive to this effect, whether it is true or false. [p. 96, italics added]

That the information needs to be close enough to the truth, as emphasized, does undercut some of the force of this as an EDA, but suppose that the fitness enhancer here is convincing others, not reaching truth, and the EDA returns. This is a form of extreme debunking that could be used to argue (ironically) that reason is not a faculty that evolved to track truth either. EDAs are ubiquitous. But to the extent that the fitness enhancers *are* truths, EDAs do not have universal effect. There remains the meta-issue of how we know that the fitness enhancer is truth. Most of the time we merely have expectations or informed guesses and merely know *that* a selective sweep has occurred, but in this chapter, I will presume that we do know.

Aspects of evolved cognition debunk religion

There are three main ways evolved cognition debunks religious beliefs: the biological, the cultural, and the sociological; the latter two are often treated as equivalent, but for my purposes, here I will treat cultural evolution, which is a process that tracks symbolic and behavioral expressions of cognitive dispositions, as distinct from sociological evolution, which tracks the functionality of institutions.[9] Each domain asks a general question of the adaptation of beliefs: *Biological*: what aspects of our evolved cognition (AECs) are involved in religion? As we have seen, none of the plausible preferred explanations of adaptive AECs track truth and indeed cannot be thought to track metaphysical truth at all by selection. *Cultural*: what AECs are in evolved in religious beliefs? We have seen a wide variety of cognitive dispositions, most of which are spandrel-type explanations. Norm enforcement is adaptive, under the watchful eye of the deity, but the adaptation here is for cooperation, not the truth of the belief a deity is watching. So cultural adaptation explanations are not truth-tracking. *Sociological*: What AECs are involved in religious social institutions? Generally, these are framed in terms of social functionality of institutions—like costly signaling (Sosis 2004). Costly signals must not be something many individual believers can converge upon; they must not therefore be beliefs that are likely to be true. Similarly, the functionality of institutions is what matters, not the post hoc justifications for agents to participate in them, and so these "false consciousness" beliefs are very unlikely to have resulted from truth-tracking.

We can therefore conclude that religious belief *qua* religious belief is debunked by EDAs in the absence of any good reason to think otherwise. That is, our confidence in the truth of religious cognition, beliefs, and institutions is either weakened or deflected to other explanations such as group fitness. If genealogical explanations are all that we rely upon for religious beliefs and practices, then we must reasonably reject religious beliefs and practices as truth-trackers, although they me be fitness enhancers.[10] At the least having a naturalistic explanation of religion must weaken our confidence in the truth of any religion to some degree: that is, to the degree that the warrant of those beliefs is gained from nonnaturalistic explanations such as revelation or mystical experiences.

Science, on the other hand, despite the presence of error and the fallibility of its practitioners, is not debunked by EDAs, as the fitness of the beliefs tracks environmental truth for common sense, and the extreme costs of error elimination are borne in science in ways that it is not in religion; or, for that matter, in common sense outside the experiential domains to which common sense adapted.

One way to conceive of this is as beliefs existing along a continuum of selective environments from natural facts to constructed facts in Figure 2.1. Assuming that our best science must in some degree be constructed, and so contained elements of social or cultural adaptation (e.g., when a theory is conceived and implemented in terms of economic ideas or applications, as Darwin's theories are thought to do by some), science is never entirely subjected to the sorts of selective pressures that would track truth fully and completely,

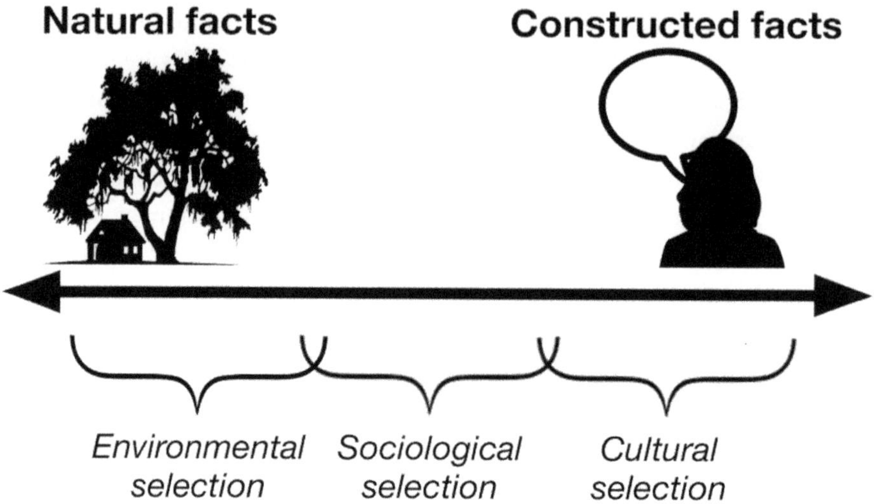

Figure 2.1 Selective pressures occur along a continuum of natural facts to socially and culturally constructed facts; environmental selection in a social organism is never entirely free of social aspects, and cultural selection at the other end of the scale is never entirely free of some natural aspects. Sociological selection, founded upon success in social behaviors, depends upon both. One might situate folk psychology, for example, at the cultural edge of sociological selection, since the folk categories of agency are strongly determined by cultural norms; while folk physics will be situated at the constructed edge of environmental selection, since some assumptions like falling speed covarying with weight are sufficiently accurate for cultural work, until physics as a social institution increases the precision and accuracy required. Religious beliefs will undergo selection in the region between purely sociological selection and purely cultural selection; but even then, facts about human psychology and neurobiology will play an important role.

even if it were true that selection was a globally optimizing process, which it is not. Nor do beliefs formed at the other end of that scale exist in a hard vacuum of the nonempirical as some postmodern theorists like to suggest.[11] Consequently, the older internalism/externalism approach to science has no more sharp boundaries or pure cases than it does in religion; all beliefs are partly a response to the natural world and partly a response to the socially and culturally constructed world. However, we can see that the success of metaphysical beliefs generally, and religious beliefs in particular, are not correlated with any selective pressures that track truth. As Griffiths and I say, this is to conflate two distinct levels of explanation. The property that makes religious beliefs fit is not their truth.

In conclusion, we can say that on the best present naturalizing accounts of religious belief and institutions, evolutionary accounts in particular and genealogical accounts in general do tend to debunk religious belief, in the sense that they undercut our warrant for accepting the content of them as true. This, by contrast, is not the case for our best science, although the two domains are not divided by any sharp qualitative boundary in this respect.

Notes

1 By "belief" here, I mean only that there are statements that are held to be true or very close to the truth.

2 The fitness can be of the beliefs themselves, as in the memetic account of Dawkins et al., or of those who on average tend to hold the beliefs, as Darwin himself thought. Again, it does not matter for our account. In the end, what counts is the spread of beliefs in virtue of their contributing to ecological success.

3 There are several accounts of social evolution, but it does not matter what account is adopted, for this argument. The fact remains that we learn about the world through a process of trial, error, and transmission, whether that transmission is dual or singular, or something else, and whether it becomes a genetic transmission process over time or not. Beliefs do evolve over time.

4 "Or perhaps he thinks the tiger is a large, friendly, cuddly pussycat and wants to pet it; but he also believes that the best way to pet it is to run away from it" (Plantinga 1993) p225.

5 Readers will recognize this as an example of the miracle argument presented by Putnam and others (Musgrave 1988), in which the continued success of science can only be explained by its truth, or rather by adopting an attitude of scientific realism. I am not suggesting that we need to go so far as full-blown scientific realism, however. All that is needed is that science tracks truth, not that it expresses it. One may be a realist on evolutionary grounds; one is not required to be.

6 There can be a common-sense view of disease based on demonic activity, for instance, so long as the epidemiology and symptomatology are right. See also Musgrave's (Musgrave 1988) point about astronomy and epicycles.

7 Much popular debate about the cognitive aspect of religion presumes that since religion is false, and clever people believe in it, they must be the subject of an illusion (a mild form of insanity). Otherwise, as Dawkins once noted ("Put Your Money on Evolution," *New York Times*, April 9, 1989, p35), you have to believe that religious believers (technically, believers in creationism) are ignorant, wicked, or stupid. Dennett's view (Dennett 2006) that religion is a pathogen that infects brains makes a similar presumption (apart from anything else, pathogens are often benign, for good evolutionary reasons, and pathogens whose replication rate

approaches that of their hosts', as religion tends to, evolve commensality rather than pathogenicity (Ewald 1994)). Hence, I shall not presume that neither religions, nor religious beliefs, are maladaptive at the outset.

8 This may raise the issue of the application of the genetic fallacy: that the source of information somehow denigrates the information. However, if there are competing explanations—such as the origin of a belief is not consistent with claims of the transcendental nature of the belief, as in this case—then while this is a genetic argument, it is not so much a fallacy as a shifting of the onus on those who believe the transcendental nature of the beliefs to show that a natural account does not remove the necessity for thinking the belief is transcendental. Of course, a genetic account does not disprove the transcendence of a belief; but it reduces or even eliminates reasonable warrant for thinking that a belief is transcendent in the absence of any other stronger argument.

9 Why the functionality of institutions matters depends upon the theoretical framework of the sociological traditional appealed to: it may be economic, as in the Marxian tradition, or power related, or even psychological. It must be obvious, but I will state it anyway, that these varieties of naturalization shade into each other: cognitive accounts depend upon human biology, cultural accounts can be given in terms of a sociological process, etc.

10 In nonnaturalistic epistemologies, this may not, of course, be the case (Baillie 1956).

11 Not all, and not only postmodernists; some Gestalt thinkers in the philosophy of science also like to pretend that theories are entirely constructs.

Bibliography

Baillie, J. 1956. *The idea of revelation in recent thought*. New York: Columbia University Press.

Barrett, J. L. 2013. Exploring religion's basement: The cognitive science of religion. In R. F. Paloutzian and C. L. Park, eds, *Handbook of the psychology of religion and spirituality*, 234–255. New York: Guilford Publications.

Barrett, J. L. and B. Malley. 2007. A cognitive typology of religious actions. *Journal of Cognition and Culture* 7: 201–211.

Boyd, R. and P. J. Richerson. 1988. The evolution of reciprocity in sizable groups. *Journal of Theoretical Biology* 132(3): 337–356.

Boyer, P. 2001. *And man creates God: religion explained*. New York: Basic Books.

Bulbulia, J. 2004. The cognitive and evolutionary psychology of religion. *Biology and Philosophy* 19(5): 655–686.

Cumming, G., F. Fidler, and D. L. Vaux. 2007. Error bars in experimental biology. *Journal of Cell Biology* 177(1): 7–11.

Darwin, C. R. 1888. *The life and letters of Charles Darwin: Including an autobiographical chapter* (F. Darwin, ed.). London: Murray.

Dawkins, R. 2006. *The God delusion*. London: Bantam Press.

Dennett, D. C. 2006. *Breaking the spell: Religion as a natural phenomenon*. London: Allen Lane.

Ewald, P. W. 1994. *Evolution of infectious disease*. Oxford [England]; New York: Oxford University Press.

Frankel, F. and R. Reid. 2008. Big data: Distilling meaning from data. *Nature* 455(7209): 30–30.

Godfrey-Smith, P. 1991. Signal, decision, action. *Journal of Philosophy* 88: 709–722.

Hacking, I. 1983. *Representing and intervening: Introductory topics in the philosophy of natural science*. Cambridge: Cambridge University Press.

Kahane, G. 2011. Evolutionary debunking arguments. *Noûs* 45(1): 103–125.

Leslie, A. M. 1987. Pretense and representation: The origins of "theory of mind." *Psychological Review* 94: 412–442.

Mackie, J. L. 1990. *Ethics: Inventing right and wrong, Penguin philosophy*. London: Penguin Books.

Mason, R. 1997. *The God of Spinoza: A philosophical study*. Cambridge; New York: Cambridge University Press.

Mercier, H. and D. Sperber. 2011. Why do humans reason? Arguments for an argumentative theory. *Behavioral and Brain Sciences* 34(2): 57–74.

Mohrhoff, U. J. 2008. A comment on Alvin Plantinga, "Evolution vs. Naturalism." *AntiMatters* 2(3): 85–86.

Musgrave, A. 1988. The ultimate argument for scientific realism. In R. Nola, ed., *Relativism and realism in science*, 229–252. Dordrecht: Kluwer.

Plantinga, A. 1993. *Warrant and proper function*. New York: Oxford University Press.

Plantinga, A. 2008. Evolution vs. naturalism: Why they are like oil and water. *AntiMatters* 2(3): 79–84.

Psillos, S. 1999. *Scientific realism: How science tracks truth*. London; New York: Routledge.

Quine, W. V. O. 1993. In praise of observation sentences. *The Journal of Philosophy* 90(3): 107–116.

Ruiz, A. C. 2013. *Evolutionary debunking arguments and their challenges to human knowledge*. Ohio: Ohio University.

Shariff, A. F. and A. Norenzayan. 2007. God is watching you: Priming God concepts increases prosocial behavior in an anonymous economic game. *Psychological Science* 18(9): 803–809.

Sosis, R. 2004. The adaptive value of religious ritual. *American Scientist* 92(2): 166–172.

Uexkull, J. v. 1926. *Theoretical biology, international library of psychology, philosophy and scientific method*. London: Kegan Paul, Trench, Trubner.

Van Fraassen, B. C. 1980. *The scientific image*. Oxford: Clarendon Press.

Wilkins, J. S. 2015. Gods above: Naturalizing religion in terms of our shared ape social dominance behavior. *Sophia* 54 (1): 77–92.

Wilkins, J. S. and P. E. Griffiths. 2013. Evolutionary debunking arguments in three domains: Fact, value, and religion. In J. Maclaurin and G. Dawes, eds, *A new science of religion*, 133–146. Chicago, IL: University of Chicago Press.

3

Cognitive Science, Evil, and God

John Teehan

Introduction

CSR brings the various disciplines and methodologies of the cognitive sciences to the investigation of religious phenomena. While disputes continue, the establishment of several consensus points allows us to speak of a general cognitive model. A foundational position of the field, at least as it relates to god-beliefs, is articulated by Justin Barrett, who writes,

> Belief in gods requires no special parts of the brain. Belief in gods requires no special mystical experiences.... Belief in gods requires no coercion or brainwashing.... Rather, belief in gods arises because of the natural functioning of completely normal mental tools working in common natural and social contexts. (Barrett 2004, 21)

CSR posits that belief in supernatural beings (e.g., gods, ghosts, spirits, demons) arises from the workings of a suite of mental tools that evolved to serve survival/reproductive needs of our earliest ancestors. These tools—for example, Agency Detection Device, Theory of Mind, Promiscuous Teleology, Common Sense Dualism—function as part of what has come to be called System 1 (Kahneman 2011). System 1 refers to mental processes that are quick, automatic, and generally outside of conscious control. Such processes result in compelling intuitions that not only shape our interpretations of the world, but in many cases constitute our very apprehension of the world. These are not unique to religious cognition but are fundamental aspects of our perceptions, emotional reactions, and moral intuitions, and cognitive science posits that they are by far the dominant aspect of our cognition. In addition to System 1 processes, there are also System 2 processes. These are slower, more deliberative, "rational" processes that we typically associate with conscious thought and reason.

System 2 processes are important and do influence behavior and belief, but they are relatively weak compared to the intuitive, emotionally valenced outputs of system 1 (Haidt 2001; Kahneman 2011). As products of System 1 processes, religious beliefs are intuitively compelling, emotionally rich, and relatively immune to rational refutation.

In addition, there is also general consensus that religion[1] has played a role in human social evolution in serving to unite groups into cohesive, functioning social units. Just how early on religion played this role, the relative importance of other social systems and whether this role worked strictly on the level of cultural evolution or also had an effect on biological evolution are issues of contention (e.g., Alcorta and Sosis 2005; Pyysiainen and Hauser 2009; Atran and Henrich 2010; Slingerland et al. 2013); that religion played a key moral function for humans and human society is not. This is significant for our discussion of religion and evil.

While an admittedly general overview of the findings of CSR, this should prove sufficient to allow us to consider some implications for religious belief.

Implications for religious belief

Evolution-based cognitive science is the foundation of a powerful naturalistic theory of religion. It provides an empirical, experimentally grounded account of the formation of the core components of god-beliefs and of the cognitive and emotional systems that make these beliefs salient and that also serve as protective measures against criticism—and all of this holds true even if there is no God.[2] As such, CSR may pose the most serious empirical challenge to religious belief to date.

Strangely, however, you might not necessarily come to this conclusion through a review of the literature. Consider the following claims from by some scholars involved in this discussion. Justin Barrett confidently reassures believers that, "even if this natural tendency toward belief in God can be conclusively demonstrated to be the work of evolved capacities, Christians need not be deterred" (2004, 123), and Barrett is not alone in this confidence. For example, it has been argued that "CSR does not show that belief in god is irrational" (Thurow 2013, 77); that "antitheistic claims about the *incompatibility* of the CSR with theism look like they may be harder to maintain than first appearances might suggest" (Leech and Visala 2011, 47); and that "the significance of these [CSR] theories for the holding of religious beliefs is not very great" (Peterson 2010, 545).

One may be forgiven for wondering just what is going on here. How could it be that a naturalized account of god-beliefs as by-products of evolutionary processes not be significant? How is it that Christians "need not be deterred" and continue to hold onto the rationality of such beliefs? Once again, we can turn to Justin Barrett to provide a succinct statement in response to this apparent quandary: there is no deep challenge because "God may have fine-tuned the cosmos to allow for life and for evolution and then orchestrated mutations and selection to produce the sort of organisms we are ... " (2004, 123), or, as biologist Kenneth Miller put it commenting on the evolution/religion debate, "Evolutionary forces become just one more tool in the hands of the Almighty" (1999, 285).

So, just as the challenge posed by Darwin's introduction of evolution via natural selection was met by viewing natural selection as guided by divine providence, with God creating through the evolutionary processes, cognitive evolution can be understood as the method God chose for bringing about creatures with the kinds of minds capable of coming to know Him. The discovery by cognitive science of god-generating mechanisms (Shults 2014) is simply the discovery of God's method of self-revelation.

This position may be considered an updated version of natural theology. In natural theology, belief in God, and beliefs about God, are framed in such a way as to be compatible with, even supported by, the findings of natural science. The notion that our cognitive structures serve as means of God's self-revelation stands in a long line of such theological and philosophical efforts, now updated to incorporate the latest, relevant empirical evidence. We shall refer to the belief in God justified in this manner as a cognitively natural theology,[3] and beliefs about God grounded in this manner will constitute a cognitively compatible Theism, or cTheism,[4] that is, a system of beliefs about God that are supported by or at least not inconsistent with, or in conflict with, the best findings of cognitive science. How successful is such a theology as a response to the challenge of cognitive evolution?

While the analogy between theology's response to biological evolution and to cognitive evolution may seem apt, there is a significant difference being overlooked. Theists may safely place God's role outside the evolutionary process because evolution has nothing to say about what goes on outside of evolution. Theists may therefore accept a wholly naturalistic account of human origins and still feel warranted in believing in a God who designed that process. They can then relegate the resolution of tensions between those two accounts to the realm of theology or philosophy. However, an evolutionary account of cognition brings belief itself into the domain of science. Where our beliefs come from, to what degree they may be reliable, truth-tracking, rational as opposed to intuitive, is to a significant degree, at least in many cases, crucially related to a cognitive analysis. Therefore, the belief that God designed the evolutionary process, as a *belief*, falls within the domain of an evolved-cognitive science. And here is the true crux of the matter: the question is not "Did God design the process of cognitive evolution to result in minds capable of coming to know him?," but rather "What reasons, in the light of CSR, do we have for believing 'God designed the process of cognitive evolution'?" This is not a question of logical compatibility, but whether we have sufficiently solid grounds for accepting that belief. Is this belief supported by, or at least consistent with, the best available evidence from cognitive science? Or does CSR "debunk" religious belief?

There is a growing literature addressing just this question (e.g., Murray 2009; Clark and Barrett 2010; Kahane 2011, Leech and Visala 2011; Barrett and Church 2013; Draper and Nichols 2013; Shults 2014; Teehan 2014), with those arguing for the debunking position, as well as those defending the plausibility of cTheism. It is not my intention to address the various facets of this debate. Instead, I plan to explore an apparently unnoticed aspect of the issue: the relevance of CSR for the problem of evil. It should, appropriately, strike one as strange to refer to the problem of evil as an unnoticed challenge,[5] given that it is, historically, one of the most significant challenges to belief in God and as such has generated voluminous theological attention. The presence of evil, and evil of such a degree and extent, in a world claimed to be the creation of an omnipotent, omnibenevolent God raises serious moral and conceptual problems. The attempt to justify God, or God's ways in the face of such apparent evil, is theodicy. The inability to develop a successful theodicy threatens to undermine belief in God and represents a failure of Theism. There is, of course, a body of literature addressing this problem, stretching back thousands of years. This is important but not relevant here. My focus is strictly on the import of CSR for this issue, and I will argue it is significant.

Cognitive science changes not only how we understand God but how we understand evil, and these changes have ramifications for belief. My thesis is that a Theism that can be made consistent with the findings of CSR—that is, that cognitive science reveals God's natural means of self-revelation—cannot also be made consistent with a cognitively based understanding of evil; that is, a cognitive science of evil provides a powerful evidential argument against cTheism.

Cognitive evolution, morality, and evil

To appreciate the special challenge of a cognitive account of evil, we must recognize a previously unstated implication of a cTheism: if God guided the process of cognitive evolution to result in the kinds of minds capable of coming to know Him, then this applies not simply to god-generating mechanisms but also to morality-generating mechanisms, as well. This entailment is particularly salient for the God of Theism, who is understood as an essentially moral Being, intimately concerned with the morality of His creatures. This connection between god-beliefs and morality is also attested to by the findings of CSR (e.g., Sosis and Alcorta 2003; Atran and Norenzayan 2004; Johnson and Kruger 2004; Sosis 2006; Shariff and Norenzayan 2007; Teehan 2010; Norenzayan 2013). Therefore, our evolved moral cognition is as much a revelation of God's nature as are our god-generating mechanisms. Given this, an understanding of an evolved-cognitive account of morality is in order.

The ability to live cooperatively in large, uniquely complex social units is perhaps the defining characteristic of the human species, but for this to be possible humans must address the tension between an individual's promoting his/her own inclusive fitness (the standard target for natural selection) and the need to contribute to the stability and effectiveness of the group—which require transferring resources from self and family to the group. Given the social nature of *hominins*, a stable and flourishing group creates the conditions under which individuals may more successfully pursue their own inclusive fitness. Therefore, resources committed to the group function as a sort of long-term investment in one's own fitness. However, this only works if others in the group also invest in the common pool, and there exists the temptation to not invest and yet benefit from the group's common resources. This failure to reciprocate (whether it occurs directly, group member to group member, or indirectly, by not investing in the group from which you benefit) increases the cost to those who do cooperate with the group. This threatens to undermine the cooperative schema which sustains the group. Morality, from an evolutionary perspective, is a system for promoting prosocial behavior and discouraging and punishing socially costly behavior. It is a means for addressing the problem of individuals pursuing their own inclusive fitness within a social setting. There is a substantial theoretical and experimental literature attesting to the evolutionary function of moral systems (e.g., Alexander 1987; Boyd and Richerson 1988; Gintis 2000; Fehr and Fischbacher 2003; Nowak and Sigmund 2005; Nowak, 2006).

An evolutionary theory of morality is quite complex, and a detailed discussion is beyond the scope of this paper. However, there is one key feature of evolved morality vital to our discussion: morality is an in-group adaptation. The group is the setting for our pursuit of inclusive fitness, where we find protection from the dangers of an uncertain environment (and from other groups), where we raise our children and care for our extended families (the source of our genetic future). It is where we invest our resources; it is the source of the return on that investment. It is in-group members who must be counted on to cooperate, reciprocate, and invest in the group; it is in-group members whose violation of the social code poses the most significant and regular threat. Those outside the group are less often, if ever, in a position to be partners in social exchange. They are not personally invested in the group and, in fact, are often threats to the group—whether in terms of potential aggression or in terms of being resource-competitors. The in-group boundary sets the boundary of moral concern. This results in a moral psychology that is sensitively attuned to signals of in-group membership, as well as indications that one's commitment to the in-group is unreliable, that is, cheating and defecting. We are, by nature, more morally sensitive to those we recognize as in-group members than to out-group members, who do not trigger the same moral intuitions and, in fact, may trigger fear and threat-responses. This in-group/out-group moral divide is the psychological source of much of the compassion, love, pride, and willingness to sacrifice that humans can demonstrate. It is also the source of xenophobia, ethnocentrism, racism, and narrow tribalism that also defines human behavior—and which constitute types of moral evil.

In assessing moral behavior, evolutionary theory tends to focus on ultimate causes, such as kin selection, indirect altruism, reputation management, etc. Ultimate causes concern evolutionary cost/benefit ratios that favor one strategy over another, all measured strictly in terms of reproductive differentials. Human behavior, however, is not typically motivated by ultimate causes. A parent's investment in her/his child, while contributing to inclusive fitness, is not necessarily motivated by calculations of genetic returns. Rather, it is the love and pride that one feels for a child that serve as proximate causes for parental investment. Humans evolved to feel love and pride in children is to be explained by reference to ultimate causes, but the translation of these ultimate causes into behavior must focus on proximate causes (Scott-Philips et al. 2011). This helps to avoid some misplaced criticisms of evolutionary theories of morality: for example, if parental investment is motivated by concern for passing on copies of one's genes, how do you explain love of, for example, adopted children, where there is no possibility of passing on copies of one's genes? The ultimate cause of parental investment has equipped us with cognitive-emotional mechanisms that can be triggered by cues that are not related to genetic reproduction, that is, by proximate causes.

In terms of moral behavior more generally, research is coming to focus on a particularly important proximate cause: empathy. Empathy is the capacity to recognize and respond to, on both a cognitive and emotional level, the emotional state of another individual (Decety and Jackson 2006; Shamay-Tsoory 2011). This capacity for empathy is instantiated on a neurological level and is the subject of a wide body of neuroscientific literature.

The empathetic systems of the brain are spread across diverse areas of the brain. Simon Baron-Cohen identifies at least ten areas involved in the various expressions of empathy (Baron-Cohen 2011). We need not concern ourselves with the specifics of the neurological picture, but two general points are needed: (1) that humans are equipped with cognitive mechanisms for empathy that function as part of System 1—our empathetic responses are bottom-up processes that are automatic and quick. When you witness a person fall and hit her head, your brain immediately makes a representation of that event which triggers neurological responses that generate an empathetic response, and (2) that empathy functions as the proximate cause of moral behavior—it plays a key role in triggering our moral intuitions.

Empirical support for the notion that a healthy, well-functioning empathetic response is central to morality comes from studies of individuals with damage to parts of their empathy systems. The most dramatic example is that of the psychopath. The clinical picture of a psychopath is an individual with damage to, or decreased responsiveness in, areas of the brain that process emotional cues. That is, psychopaths have a deficient capacity for emotional empathy. They may be quite capable of cognitive empathy (i.e., recognizing that someone else is in pain) but without experiencing the accompanying emotional empathy (i.e., feeling emotional distress at another's pain). This particular profile is what allows some individuals with psychopathy to engage in the cruel behavior often associated with this disorder. They do not feel emotional distress at the sufferings of others, nor do they have the appropriate fear response to the prospect of causing others to suffer. While very few individuals with psychopathy commit violent crimes, a disproportionate percentage of those who commit the most serious of violent crimes have psychopathy—and almost all serial killers appear to be psychopaths (Blair et al. 2005; Baron-Cohen, 2011).

Psychopathy is not the only example of inhibited empathetic responses being implicated in antisocial and criminal behavior. Neuropsychologist Adrian Raine has compiled an array of data connecting decreased volume in the medial ventral prefrontal cortex (mVPFC), which is involved in inhibiting impulsive behavior and homicide (Raine 2013). Neuroscientist James Blair and colleagues (2005) has established a connection between deficient amygdala responses (the amygdala plays a complex role in emotional reaction, particularly fear responses) and childhood behavioral disorders.

In terms of understanding evil, in this case the horrible and undeserved suffering humans can impose on others, cognitive science allows us to locate the genesis of much evil in the workings (or failed workings) of the human brain—a brain that cTheism claims is designed to function as a means of revealing God's nature. In response, the cTheist may confidently point out that the above examples are all the result of the empathetic systems of the brain not functioning properly, that is, not working as God designed them to work, and so, no culpability may be attributed to God. Of course, one might question a divinely guided cognitive design that is capable of going so terribly wrong, resulting in such suffering. It also raises serious questions about the moral responsibility of psychopathic individuals, and how that can be reconciled with a theology of salvation, as individuals are not responsible for having psychopathic brains. But even if we leave such concerns aside, even if psychopathological evil does not challenge cTheism, psychopathy is not the only source of evil. What of the evil humans do that is not based in a brain disorder?

The cognitive bases of moral evil

From the picture we have sketched above, we can set out an evolved cognitive account of evil. Empathy, which is the proximate cause of moral behavior, has a neurological basis that operates automatically and outside of conscious control. It is a bottom-up process, characteristic of System 1; and this process can be disrupted or inhibited by damage to any of the component neurological structures. But, research indicates that this bottom-up process can also be modulated by top-down cultural and experiential events. For example, hearing one's child cry triggers an empathetic response that motivates protective behavior in adults. Learning (a top-down process) that the child is crying in anticipation of, say, an imminent inoculation can modulate that response such that the emotional empathy does not provoke protective behavior. The extent of our empathetic responses can also be expanded by top-down processes. Learning of the cognitive complexity of, for example, dolphins, can prime us for increased moral concern for creatures that might not have previously elicited empathy. These learned empathetic responses can come to be automatic System 1 responses and so part of our repertoire of intuitive moral responses.

However, it turns out that not all top-down processes are equal. The brain seems primed to respond to evolved moral concerns: to cheaters, to cooperators, and, significantly, to in-group/out-group distinctions. This convergence between evolutionary and neurological models of morality should not be surprising. Evolutionary theory explains the selection pressures that give shape to our empathetic responses; neuroscience reveals the mechanisms by which the brain represents social relationships and motivates altruistic behavior. It makes sound evolutionary sense that we are particularly attuned to a conspecific experiencing pain or fear: their experience may signal possible threats in the environment, and so empathy may function as a threat defense mechanism (Lamm et al. 2007). Empathy, however, does more than simply alert us to self-directed dangers, it cues us to the distress of others in a way that generates helping behavior, and evolutionary theory makes very specific claims on this topic: it will be primarily directed toward members of the in-group. There is a substantial body of research that fills in the details of this convergence.[6]

Studies by Tania Singer and colleagues reveal intriguing neural responses to cooperators and cheaters. In one experiment, participants played a Prisoner's Dilemma game with confederates of the experimenters, who were instructed to either cooperate or cheat. The subjects were then scanned in an fMRI while viewing images of the confederates experiencing a painful stimulus. Both male and female participants exhibited empathetic responses to those confederates who had cooperated during the game but had reduced empathetic brain responses to the confederates who had cheated. Not only were there reduced empathetic responses, the male participants exhibited increased activation in the reward centers of the brain (Singer 2006; Singer et al. 2006).

Perhaps more significant for our purposes is evidence of a neurologically based bias against out-group members. In a widely cited study, Xiaojing Xu and colleagues (Xu et al. 2009) compared the neural empathetic responses to seeing others receive painful stimuli. The subjects consisted of Chinese and Caucasian participants, who were shown video clips of other Chinese and Caucasian individuals receiving the stimuli. The results showed that witnessing the painful stimuli increased activity in brain areas involved in first-person pain experiences (i.e., the anterior cingulate cortex, and the frontal/insula cortex), and this was independent of the race of the participant or the person in pain. However, these empathetic responses "were greater to racial in-group than out-group members," and this held true of both Caucasian and Chinese participants (Xu et al. 2009, 8527). Interestingly, this bias was not found in explicit assessments of degree of painfulness. That is, when asked to rate how much pain the person in the video clip was experiencing, no racial bias was detected. This indicates that humans have an automatic, neurological empathetic response to the suffering of other humans, but that this base response can be modulated by in-group status.

The biasing effect of racial in-groups has been found in numerous studies (e.g., Phelps et al. 2000; Richeson et al. 2003; Avenanti et al. 2010; Chiao and Mathur 2010), but it is not only racial groups that have this effect, cultural in-groups have also been shown to modulate how the brain responds to the pain of others, the recognition of emotions in others and amygdala responses to fear faces (Elfenbein and Ambady 2002; Chiao et al. 2008; Han and Northoff, 2008; Mathur et al. 2010). In fact, one study demonstrated an in-group bias to even "minimal groups"—that is, artificial groups organized for a particular purpose—in terms of facial recognition (Van Bavel et al. 2008). There is mounting neural evidence that we are implicitly biased toward members of our in-group in ways that are significant for morality: we have a moral bias toward in-groups members, along with a decreased sensitivity to out-group members (Elfenbein and Ambady 2002; de Waal 2008; Van Bavel et al. 2008; Chiao and Mathur 2010; Mathur et al. 2010).

And it is not simply decreased sensitivity. Certain signals of being in the out-group can actually trigger an ingrained disgust response. In a study by Harris and Fiske (2006), images of low-status, out-group members failed to trigger those neural systems involved in processing persons and instead triggered those systems that respond to disgusting objects. These people did not even register in the brain as persons. They were objects of disgust, and objects of disgust are to be avoided or eliminated.

The evolved moral bias against the out-group, a bias that is neurologically instantiated, is not the source of all moral evil, but it is the psychological basis of prejudice, discrimination, and dehumanization—and all the injustice, harm, and violence that follow. Significantly, this results not from damaged or deficient brains; this all follows from the brain *working as it was designed to work*. This leads to an empirical conclusion: moral evil (at least some of the worst examples) flows from the workings of mental tools that evolved to promote inclusive fitness and in-group fitness. It also allows us to draw a moral conclusion: If

God designed our cognitive evolution as the means to revealing himself to our minds, then these cognitively based evils are predictable, even inevitable, results of God's moral design. The suffering produced is not simply collateral damage to the fulfillment of some otherwise worthy moral goal; the establishment of a moral bias against other groups of humans is a design feature of our moral minds. As God knew this would be the case, God bears responsibility for moral evil. If God bears responsibility for moral evil—if the degrading, destructive ways humans treat other humans, for selfish or tribal reasons, is a natural result of divine design—then, I propose, theodicy fails.

If a cognitive understanding of evil makes the case for God's culpability for evil, then theism cannot both (a) be made compatible with the facts revealed by an evolved-cognitive science (i.e., argue that God reveals himself through our cognitive architecture) and (b) defend God in the face of the evidential argument presented by cognitive science. Therefore, an evolved-cognitive science of evil threatens to undermine cTheism.

This provides powerful evidence supporting the case for debunking. Still, as this is a relatively unexplored challenge, perhaps cTheism can establish a cognitively compatible theodicy (cTheodicy) in response. I am not aware of any such efforts, which does not mean one cannot be developed. There are suggestions in the literature defending cTheism that may be relevant. To avoid too premature a conclusion on cTheism, let us try to construct such a theodicy.

cTheodicy

In line with the cognitive-natural theology we are working with, we might hypothesize that God did indeed endow us with a moral cognition that would prepare us to act in the manner consistent with the command to love our neighbor as ourselves and "to do justice, to love kindness, and to walk humbly" (Micah 6:8), but that something went wrong. Barrett suggests perhaps "a perfectly adequate concept of God does come as part of our biological heritage but that living in a sinful, fallen world this concept grows corrupt" (2009, 97). Drawing on Christian theology, Alvin Plantinga points to the disastrous "cognitive consequences" (Plantinga 2000, 205) of original sin, that left us epistemically corrupt. Perhaps it left us morally corrupt as well. Indeed, this reading fits readily with a traditional theology of original sin, that is, that pride, the original sin, leads us to prioritize our will and good over that of God and certainly over that of others. This prioritizing of our good gives rise to the in-group bias that we have presented as the source of so much evil. This moral corruption, then, needs a moral-cognitive fix, which God provides through revelation. Evil is a result of a deviation from God's design, not a natural outcome of that design. Thus, God is justified. How does this cTheodicy fare?

This approach is difficult to reconcile with natural theology. One central problem is the concept of original sin seems inconsistent with natural history. If we are to read Genesis as even a roughly historical account, with an original pair of humans committing that first sin, then there is no possibility of reconciliation with a naturalistic account that sees humans as part of an evolutionary tree, with modern humans arriving very late in the process. Of course, Genesis need not be read that way and is not read that way by many Theists. "Adam and Eve" may be metaphors for the first morally or spiritually mature members of *homo sapiens*, and so the first truly human beings made in the likeness of God (e.g., Korsmeyer 1998; Williams 2001). Original sin, in this evolutionary context, is not the first sin committed by the original humans but instead is an event that took place in the history of the species, once humans became moral agents. In terms of a theology compatible with science, this is a better fit—but is it a good fit?

Even this naturalistic update to the doctrine of original sin entails that there was a moral fall, a moment in the development of the species that constitutes a break with a previous moral condition—and this is incompatible with what we know of our evolutionary history. There was no pristine moral state from which to fall. The moral evils we are considering, those that follow from an in-group/out-group bias, are found throughout the higher mammals. This speaks to a deep evolutionary origin of the in-group bias—and all the moral insensitivity and cruelty that it generates—one that predates the human-chimp split, some six million years ago. Just how far back must we go to find this Fall? (Hick 2001; De Cruz and De Smedt 2013).

Perhaps, however, one might want to argue that the discriminatory urgings of the cognitively based in-group bias are not themselves the source of sin but are primitive elements of the evolutionary ascent to truly moral persons; that the process of ensoulment, which creates the "first humans," endowed us with a moral perspective that allows us to resist the myopic morality of our evolved moral psychology and instead embrace the more inclusive love that God expects of us. It is the failure to embrace this perspective and instead pridefully focus on the self and in-group that constitutes original sin. This may seem to fit better with natural theology, but not with cNatural Theology. We just have no evidence of any period in human history in which we find an inclusive love for the stranger that (a) breaks with a previous state of in-group bias common to other mammals and b) then gives way again to rampant in-group bias, ostensibly resulting from original sin. Original sin and the Fall have no place in natural history; and in terms of cognitive science, the evidence is clear that we have always been naturally inclined to in-group bias.

The theology of original sin is ill-suited to serve as part of a cTheodicy and so cannot salvage cTheism. However, original sin has always been a theologically complicated notion, even before the advent of evolutionary science. Whether or not contemporary theologians can come to a coherent understanding compatible with our evolutionary history remains a live issue for many. I am deeply skeptical that such an account can be developed, but it is not necessary to settle the issue here. For even if we were to grant this point, for the sake of the argument, and use original sin as part of a cTheodicy, serious problems remain.

According to theorists taking this approach, original sin corrupts our cognition thus leading to the moral evil we find around us, but God, in his mercy, provides us with a fix to this corrupted cognitive program (Plantinga 2000, 206) through revelation, as expressed, for example, in Scripture. By following God's moral guidance, we can correct the corrupted outputs of our moral cognition and avoid the evil that follows. Once again, God is justified; but once again, there is a problem: Scripture does not fix our moral in-group bias; it gives divine sanction to that bias.

As I have argued elsewhere (Teehan 2010), religious texts set out a moral system that follows the contours of our evolved psychology and thus endorse the in-group bias. We do not have the space to exemplify this claim in detail, and so some representative examples will have to suffice. Let us look at the commandment, "You shall not murder" (Exodus 20:13). This is a good candidate for a universalist morality, as it seems a categorical imperative unconstrained by group membership, but that is not the way it functions within the Hebrew Bible. This command clearly is not applicable to out-group members:

> When the Lord your God gives it [a foreign town] into your hand you shall put all its males to the sword, but the women and little ones, the cattle, and everything else in the city, all its spoil, you shall take as booty for yourselves; and you shall enjoy the spoil of your enemies. (Deut. 20, 13–16)

Here it is clear that the humans living in a foreign town, that is, out-group members, do not enjoy the protection of "You shall not murder." Nor are they accorded status as full-fledged human beings. The women and the children may be treated as booty—as dehumanizing an attitude toward the other as can be imagined. What sort of moral system allows humans to use children as things to be enjoyed as the spoils of war? A moral system that is designed to promote the inclusive fitness of the individual and the in-group. Rather than correcting an ostensible moral corruption of God's plan, this passage sanctifies it. And of course this is not the only example. Deuteronomy portrays the march of the Israelites to the Promised Land as a genocidal bloodbath: "We captured all the towns—there was no citadel that we did not take from them—sixty towns.... And we utterly destroyed them ... in each city utterly destroying men, women, and children" (Deut. 3:4-6).

As much as some Christians might want to characterize this moral discrimination as a concession made to the Jews that was superseded by the teachings of Jesus, we find the same moral psychology at work in the Gospels, and indeed throughout the history of Christian thought, albeit at times portrayed in less dramatic ways. Christianity redefined the boundary of the moral in-group, moving away from ethnicity as a signal of membership and establishing in its place belief in Christ. Yet, this new definition of in-group membership brings with it the moral bias that follows from this psychology. Christ himself is said to have proclaimed: "Whoever is not with me is against me, and whoever does not gather with me scatters" (Luke 11:23); that "everyone therefore who acknowledges me before others, I will also acknowledge before my Father in heaven; but whoever denies me before others, I also will deny before my Father in heaven." (Matt. 10:32-33). In the parable of the sheep and goats, he teaches "the King will say to those at his right hand, 'Come, O blessed of my Father, inherit the kingdom prepared for you' ... to those at his left hand, 'Depart from me, you cursed, into the eternal fire prepared for the devil and his angels'" (Matthew, 25:32-41).

There is an in-group and an out-group and woe to those on the wrong side of that divide:

> Just as the weeds are collected and burned up with fire, so will it be at the end of the age. The Son of Man will send his angels, and they will collect out of his kingdom all causes of sin and all evildoers, and they will throw them into the furnace of fire, where there will be weeping and gnashing of teeth. The righteous will shine like the sun in the kingdom of their Father. Let anyone with ears listen! (Matt. 13: 40–43)

Those who run afoul of the moral boundaries, whether by being part of another group or failing to abide by the in-group's social contract, lose their full moral status. Here Jesus characterizes them not as humans but as weeds, that is, things to be collected and discarded, subject to terrible pain and suffering—an attitude that has been played out through the sad history of religious prejudice and violence, a history not restricted to Christianity.

This is an admittedly one-sided look at the moral story in the Scriptures, which also contain powerful, inspiring exhortations to principles of justice and love. These move us to consider the humanity of the other and serve to expand the moral community, enriching our moral imagination. They serve as antidotes to the morally corrosive aspects of our evolved cognitive tools—just the role Scripture is assigned in the cTheodicy we are considering. However, Scripture also presents, under divine sanction, a moral bias against the out-group that promotes and even provokes (Bushman et al. 2007) moral insensitivity and outright cruelty (including slavery, genocide and torture)—the very sorts of moral evil they were supposed to correct.

This aspect of Scripture, and it is no small aspect, is inconsistent with the claims of this version of cTheodicy, and it leaves cTheism subject to the challenges posed by cognitive science: Moral evil flows from the workings of our evolved cognitive tools—tools designed to reveal God's nature, working as God designed them to—thus implicating God in the evil that humans do. Instead of countering the moral biases that lead to so much evil, Holy Scripture, which reveals a more refined understanding of God's will, gives those biases divine sanction. Rather than getting God off the hook for the evil that our divinely designed minds generate, Scripture reinforces God's moral culpability for all the evil that follows from that design. cTheodicy, then, fails; and if cTheodicy fails, then the prospect for defending a cTheism is severely, if not fatally, challenged.

But let us hold off once more before resting with this conclusion. Perhaps, one might argue, it claims too much, undermining one argument for cTheism but leaving alternative versions viable. Interestingly, such an alternative is raised in recent literature on this topic, a version based not in an Augustinian theology of original sin but one which goes back to an earlier Church Father, Irenaeus (Second century C.E.).

An Irenaean cTheodicy

In Irenaeus's thought, humanity's moral failings do not result from a discrete original sin that had species-wide consequences but rather are a result of God having created us spiritually immature. We do not have a corrupt nature that needs to be fixed, but an immature nature that needs to be developed. Philosopher John Hick sees this as compatible with a naturalistic account of humanity, a compatibility that "hinges upon the creation of humankind through the evolutionary process as an immature creature living in a challenging and therefore person-making world" (Hick 2001, 39). This notion of humans as immature creatures, still in the process of "person-making," has been proposed as a response to the Unreliability Argument, that is, that our cognitive tools when properly working are unreliable in the domain of god-beliefs; they do not reliably result in god-beliefs consistent with Theism. In this alternative theology, our mental tools are not unreliable: "They are simply spiritually imprecise or coarse-grained. Perhaps the function of the god-faculty is simply to make humans aware of the broad divine/moral dimension of reality" (Clark and Barrett 2011, 665).

God-generating cognitive tools are reliable in producing god-beliefs in a general, imprecise way. They are designed to make us aware of a spiritual aspect to reality. The natural, intuitively compelling sense of the divine that comes from these tools is then to be refined through Scripture and theological reflection. God designed evolution to allow for spiritual development, rather than determining fully formed beliefs that would automatically present themselves to us—a process that allows us to meaningfully participate in God's creation. This may be a better, more refined understanding of how God intended our cognitive tools to work, opposed to a view that just expects them to generate irresistible truth. It is an interesting approach that merits consideration (see, e.g., Clark and Barrett 2010, 2011; De Cruz and De Smedt 2013) and may provide a promising model for cTheism. But what about evil? What kind of theodicy works with this model?

John Hick presents one such possibility: moral evil is an inevitable consequence of there being spiritually/morally immature creatures, such as humans. The justification of there being such immature, evil-prone creatures is the greater good to be achieved through fulfilling God's Purpose, which is "the full realization of human potential in a spiritual and moral perfection within the divine kingdom" (Hick 2001, 44). Unfortunately, the suffering-mass of humanity rarely reaches this spiritual/moral perfection during their stay on this "person-making world" and so further theological speculations are required of Hick to make this all work out. But let us leave that concern aside for now and focus on an assessment of this as a cTheodicy.

How is our immature, unrefined spiritual nature to be drawn closer to the Truth? Certainly within the Abrahamic traditions, Scripture serves that role.

However, as argued, while Scripture certainly moves us away from a narrow individual selfishness, structured as it is by our evolved moral cognition, it does so by moving us to embrace a corporate selfishness that places the in-group, however broadly defined, at the center of our moral concerns. This is, perhaps, an improvement but certainly not a "full realization" of "moral perfection." This critique, however, may not be as problematic as it first appears, as many theologians assert that Scripture needs to be understood in more sophisticated ways. Examples of apparently God-sanctioned evil are not to be read as final revealed truth. The reading of Scripture, which serves to refine our rough sense of the divine, is itself in need of refinement via theological reflection—and this seems reasonable. But, cTheists should be wary of this move, for it opens up a slippery slope that may ultimately lead to an even more damaging conclusion.

Consider the argument: Spiritual awareness is imprecise and immature and so requires development and refinement to reach True Belief. This process of development, however, seems to be an ongoing process. Our intuitive beliefs are refined by Scripture, but our Scriptural beliefs need to be refined by theology—and what about theology? We have already considered the move to refine the theology of original sin from a naturalistically incompatible Augustinian model to a more compatible Irenaean model. But, scholars in CSR exploring the Irenaean model recognize that it too needs to further refinement, via continued theological reflection. At what point does this process of ongoing refinement stop? While it may be possible to identify immature beliefs to be discarded (although, see current revising of Augustinian theology in favor of formerly discarded Irenaean theology) it does not seem possible to identify the final stage of appropriately refined, mature theology (e.g., Irenaean theology giving way to … ?). If there is not a clear, objective standard by which we can determine the end of the process, then any theological system can only be accepted provisionally, open to future refinement or possibly rejection—and this includes the belief in a personal, Supreme Being, that is, the God of Theism. There are no compelling reasons to accept Theism as a final stage. It may just be one more step upon the path, destined to join animism and polytheism as primitive attempts to make sense of our world.

This leads to some deeply challenging conclusions about Theism in light of cognitive science. A cognitive scientific model of morality reveals the propensity to evil is built deep into our nature. This is not the result of some fall from grace but is a product of the cognitive processes that constitute our moral psychology. A naturalistic account explains this as the result of selection pressures favoring those cognitive predispositions that promoted the inclusive fitness of social beings. Such beings need to secure the cooperative groups in which they live and protect themselves from threats from competitive groups and individuals. These goals entail different moral sensitivities and so will result in different moral treatment for in-groups and out-groups. This account fits the available evidence much better than a Theistic account that sees our moral cognition as the design of a God seeking to enable humans to follow the demands of a higher morality. No Theism can claim to be compatible with the findings of CSR without addressing this moral challenge, and meeting this challenge may call for a very different approach.

A conception of God consistent with, much less supported by, an evolutionary cognitive science will need to be as dynamic and open to ongoing development and refinement as our cognitive processes are. In such a theism, scripture and theology must also be subject to this evolving understanding, constantly subject to revision and perhaps rejection. It may indeed be possible to develop such a cognitively compatible understanding of God, but the more successful that effort is, the less that understanding will resemble the God of classical Theism.

Notes

1 "Religion" is a contested term among scholars. Here I stipulate that by "religion" is intended a communally shared set of beliefs and practices involving one or more supernatural beings. This definition should serve to denote the subject matter as it is treated by both cognitive scientists and theologians working on the present topic and define the subject in a manner recognizable on a more general level.

2 "God" is also an ambiguous term. Here I stipulate "God" to mean a conception of the deity as the Supreme Being; a personal God who is creator of the universe, morally interested in humanity and possessing superlative qualities. I do not claim that CSR has nothing to say about other god-concepts, I believe it does, but here the focus is on a conceptualization of God that has played, ambiguities notwithstanding, a central role in human history, and which is almost uniformly assumed by those involved in debates such as this one.

3 See De Cruz and De Smedt (2015), for an important study of natural theology and CSR.

4 I will use "Theism" capitalized to indicate a rationally grounded and defensible system of belief in God, with "God" understood as stipulated above. This is to distinguish it from a more general notion of theism as any belief system about God.

5 I am not aware of research directly addressing this issue. There are relevant discussions in Draper (2004); Shults (2012); De Cruz and De Smedt (2013, 2015); Teehan (2013).

6 This discussion of the neuroscience of in-group bias is a rehearsing of the topic found in Teehan, 2012.

Bibliography

Alcorta, C. and R. Sosis. 2005. Ritual, emotion, and sacred symbols: The evolution of religion as an adaptive complex. *Human Nature* 16(4): 323–359.

Alexander, R. 1987. *The biology of moral systems: Foundations of human behavior*. New York, NY: Aldine de Gruyter.

Atran, S. and J. Henrich. 2010. The evolution of religion. *Biological Theory* 5(1): 18–30.

Atran, S. and A. Norenzayan. 2004. Religion's evolutionary landscape: Counterintuition, commitment, compassion, communion. *Behavioral and Brain Sciences* 27: 713–730.

Avenanti, A., A. Sirigu, and S. Aglioti. 2010. Racial bias reduces empathic sensorimotor resonance with other-race pain. *Current Biology* 20(11): 1018–1022.

Baron-Cohen, S. 2011. *The science of evil: On empathy and the origins of cruelty*. New York, NY: Basic Books.

Barrett, J. 2004. *Why would anyone believe in god*? New York, NY: AltaMira Press.

Barrett, J. 2009. Cognitive science, religion, and theology. In Jeffrey Schloss and Michael Murray, eds, *The believing primate*, 76–99. Oxford University Press.

Barrett, J. and I. Church. 2013. Should CSR give atheists epistemic assurance? On beer-goggles, BFFs, and skepticism regarding religious beliefs. *The Monist* 96(3): 311–324.

Blair, J., D. Mitchell, and K. Blair. 2005. *The psychopath: Emotion and the brain*. Blackwell.

Boyd, R. and P. Richerson. 1988. The evolution of reciprocity in sizable groups. *Journal of Theoretical Biology* 132: 337–356.

Bushman, B. J., R. Ridge, E. Das, C. Key, and G. Busath. 2007. When God sanctions killing: Effects of Scriptural violence on aggression. *Psychological Science* 18(3): 204–207.

Chiao, J. Y., T. Iidaka, H. L. Gordon, J. Nogawa, M. Bar, E. Aminoff, N. Sadato, and N. Ambady. 2008. Cultural specificity in amygdala response to fear faces. *Journal of Cognitive Neuroscience*, 20(12): 2167–2174.

Chiao, J. Y. and V. A. Mathur. 2010. Intergroup empathy: How does race affect empathic neural responses? *Current Biology* 20(11): 478–480.

Clark, K. J. and J. Barrett. 2010. Reformed epistemology and cognitive science of religion. *Faith and Philosophy* 27(2): 174–189.

Clark, K. J. and J. Barrett. 2011. Reidian religious epistemology and the cognitive science of religion. *Journal of the American Academy of Religion* 79(3): 639–675.

Decety, J. and P. L. Jackson. 2006. A social-neuroscience perspective on empathy. *Current Directions in Psychological Science* 15(2): 54–58.

De Cruz, H. and J. De Smedt. 2013. Reformed and evolutionary epistemology and the noetic effects of sin. *International Journal of the Philosophy of Religion* 74: 49–66.

De Cruz, H. and J. De Smedt. 2015. *A natural history of natural theology: The cognitive science of theology and philosophy of religion*. Cambridge, MA: MIT Press.

De Waal, F. 2008. Putting the altruism back into altruism: The evolution of empathy. *Annual Review of Psychology* 59: 279–300.

Draper, P. 2004. Cosmic fine-tuning and terrestrial suffering: Parallel problems for naturalism and theism. *American Philosophical Quarterly*, 41(4): 311–321.

Draper, P. and R. Nichols. 2013. Diagnosing bias in philosophy of religion. *The Monist* 96(3): 420–446.

Elfenbein, H. A. and N. Ambady. 2002. Is there an in-group advantage in emotion recognition? *Psychological Bulletin* 128(2): 243–249.

Fehr, E. and U. Fischbacher. 2003. The nature of human altruism. *Nature* 425: 785–791.

Fehr, E. and S. Gachter. 2002. Altruistic punishment in humans. *Nature* 415(Jan.): 137–140.

Gintis, H. 2000. Strong reciprocity and human sociality. *Journal of Theoretical Biology* 206: 169–179.

Haidt, J. 2001. The emotional dog and its rational tail: A social intuitionist approach to moral judgment. *Psychological Review* 108: 814–834.

Han, S. and G. Northoff. 2008. Cultural-sensitive neural substrates of human cognition: A transcultural neuroimaging approach. *Nature Reviews Neuroscience* 9: 646–654.

Harris, L. and S. Fiske. 2006. Dehumanizing the lowest of the low: Neuroimaging responses to extreme out-groups. *Psychological Science* 17, (10): 847–853.

Hick, J. 2001. An Ireneaen theodicy. In Stephen Davis, ed., *Encountering evil, a new edition: Live options in theodicy*, 38–52. Louisville, KY: Westminster John Knox Press.

Johnson, D. and O. Kruger. 2004. The God of wrath: Supernatural punishment and the evolution of cooperation. *Political Theology* 5(2): 159–176.

Kahane, G. 2011. Evolutionary debunking arguments. *Nous* 45(1): 103–125.

Kahneman, D. 2011. *Thinking: Fast and slow*. New York, NY: Farrar, Strauss, and Giroux.

Korsmeyer, J. D. 1998. *Evolution and Eden: Balancing original sin and contemporary science*. Paulist Press.

Lamm, C., C. D. Batson and J. Decety. 2007. The neural substrate of human empathy: Effects of perspective-taking and cognitive appraisal. *Journal of Cognitive Neuroscience* 19(1): 42–58.

Leech, D. and A. Visala. 2011. The cognitive science of religion: Implications for theism? *Zygon* 46(1): 47–64.

Mathur, V. A., T. Harada, T. Lipke, Y. Joan, and Y. J. Y. Chiao. 2010. Neural basis of extraordinary empathy and altruistic motivation. *NeuroImage* 51: 1468–1475.

Miller, K. 1999. *Finding Darwin's God: A scientist's search for the common ground between god and evolution*. New York, NY: Cliff Street Books/HarperCollins.

Murray, M. 2009. Scientific explanations of religion and the justification of religious belief. In J. Schloss and M. Murray, eds, *The believing primate,* 168–199. Oxford: Oxford University Press.

Neuberg, S., C. Warner, S. Mistler, A. Berlin, E. Hill, J. Johnson, G. Filip-Crawford, R. Millsap, G. Thomas, M. Winkelman, B. Broome, T. Taylor, and J. Schober. 2014. Religion and intergroup conflict: Findings from the Global Group Relations Project. *Psychological Science* 25(1): 198–206.

Norenzayan, A. 2013. *Big gods: How religion transformed cooperation and conflict.* Princeton, NJ: Princeton University Press.

Nowak, M. A. 2006. Five rules for the evolution of cooperation. *Science* 314: 1560–1563.

Nowak, M. and K. Sigmund. 2005. The evolution of indirect reciprocity. *Nature* 437: 1291–1298.

Peterson, G. S. 2010. Are evolutionary/cognitive theories of religion relevant for philosophy of religion? *Zygon* 45(3): 545–557.

Phelps, E. A., K. J. O'Connor, W. A. Cunningham, E. S. Funayama, C. J. Gatenby, J. C. Gore, and M. R. Banaji. 2000. Performance on indirect measures of race evaluation predicts amygdala activation. *Journal of Cognitive Neuroscience* 12(5): 729–738.

Plantinga, A. 2000. *Warranted christian belief.* New York, NY: Oxford University Press.

Pyysiainen, I. and M. Hauser. 2009. The origins of religions: Evolved adaptation or by-product? *Trends in Cognitive Sciences* 14(3): 104–108.

Raine, A. 2013. *The anatomy of violence: The biological roots of crime.* New York, NY: Vintage.

Richeson, J. A., A. A. Baird, H. L. Gordon, T. F. Heatherton, C. L. Wyland, S. Trawalter and J. N. Shelton. 2003. An fMRI investigation of the impact of interracial contact on executive function. *Nature Reviews Neuroscience* 6(12): 1323–1328.

Scott-Philips, T. C., T. Dickins, and S. West. 2011. Evolutionary theory and the ultimate–proximate distinction in the human behavioral sciences. *Perspectives on Psychological Science* 6(1): 38–47.

Shamay-Tsoory, S. 2011. The neural bases for empathy. *The Neuroscientist* 17(1): 18–24.

Shariff, A. and A. Norenzayan. 2007. God is watching you: Priming God concepts increase prosocial behavior in an anonymous economic game. *Psychological Science* 18(9): 803–809.

Shults, F. L. 2012. The problem of good (and evil): Arguing about axiological conditions in science and religion. In W. Wildman and P. McNamara, eds, *Science and the world's religions. Vol. I: Origins and destinies*, 39–68. Westwood, CT: Praeger.

Shults, F. L. 2014. *Theology after the Birth of God: Atheist conceptions in cognition and culture.* New York, NY: Palgrave-Macmillan.

Singer, T. 2006. The neuronal basis for empathy and fairness. *Empathy and Fairness: Novartis Symposium* 278: 20–30.

Singer, T., B. Seymour, J. P. O'Doherty, K. E. Stephan, R. J. Dolan and C. D. Frith. 2006. Empathic neural responses are modulated by the perceived fairness of others. *Nature* 439(26), 466–469.

Slingerland, E., J. Henrich, and A. Norenzayan. 2013. The evolution of prosocial religions. In P. Richerson and M. Christiansen, eds, *Cultural evolution: Society, technology, language, and religion*, 335–348. Cambridge, MA: MIT Press.

Sosis, R. 2006. Religious behaviors, badges, and bans: Signaling theory and the evolution of religion. In Patrick McNamara, ed., *Where god and science meet.* Westwood, 61–86, CT: Praeger Publishers.

Sosis, R. and C. Alcorta. 2003. Signaling, solidarity, and the sacred: The evolution of religious behavior. *Evolutionary Anthropology* 12: 264–274.

Teehan, J. 2010. *In the Name of God: The evolutionary origins of religious ethics and violence.* Malden, MA and Oxford: Wiley-Blackwell.

Teehan, J. 2013. The cognitive bases of the problem of evil. *The Monist* 96(3): 325–348.

Teehan, J. 2014. Cognitive science and the limits of theology. In R. Trigg and J. Barrett, eds, *The roots of religion: Exploring the cognitive science of religion*, 167–188. Surrey, UK: Ashgate Publishing.

Thurow, J. C. 2013. Does cognitive science show belief in god to be irrational? The epistemic consequences of the cognitive science of religion. *International Journal of the Philosophy of Religion* 74: 77–98.

Van Bavel, J. J., D. J. Packer, and W. A. Cunningham. 2008. The neural substrates of in-group bias. *Psychological Science* 19(11): 1131–1139.

Williams, P. A. 2001. *Doing without Adam and Eve: Sociobiology and original sin.* Minneapolis, MN: Fortress Press.

Xu, X., X. Zuo, X. Wang, and S. Han. 2009. Do you feel my pain? Racial group membership modulates empathic neural responses. *The Journal of Neuroscience* 29(26): 8525–8529.

4

The Explanatory Challenge of Religious Diversity

Jason Marsh and Jon Marsh

Introduction

Confronted with the world's many religious traditions, many people have felt pressed to ask the following question: how can one know that any particular religious outlook is true, given the existence of so many apparently conflicting religious outlooks? This way of interpreting the challenge of religious diversity is extremely common. It captivated Joseph Smith, for instance, and helped to spark the Latter Day Saints movement in nineteenth-century America. Troubled by the seemingly endless debate between different religious groups, Smith asked: "What is to be done? Who of all these parties are right; or, are they all wrong together? If any one of them be right, which is it, and how shall I know it?" (1902, 4).[1]

Smith's questions, though well known outside of the academy, have also motivated the main philosophical challenges from religious diversity. For instance, there is the contingency challenge, which has been said to undermine religious knowledge. As John Hick once put it, "someone born into a devout Muslim family in Pakistan is very likely to be a Muslim, someone born into a devout Hindu family in India to be a Hindu, someone born into a devout Christian family in Spain or Mexico to be a Catholic Christian; and so on (1997, 281)." There is also the challenge from peer disagreement, according to which the apparent reality of epistemic peers (i.e., equally honest and capable seekers of truth) from diverse faith traditions supports religious skepticism (Feldman 2007). Finally, there is the probabilistic challenge from diversity. According to this challenge, even if one particular religious tradition could be shown to be more probable than any of its competitors, in a one by one comparison, the combined probability of the competitors will still render it improbable, overall (Schellenberg 1997).

Such challenges can induce anxiety in believers, to be sure. But there is a growing sense among epistemologists that they all fail, in part because they rest on principles that, if endorsed, would create skeptical worries for many nonreligious beliefs (Plantinga 2000, ch 13; Kelly 2005; Bogardus 2013). Although we shall not rehearse these criticisms in detail here, many of our nonreligious beliefs are contingently held, rejected by apparent peers, and probabilistically challenged.[2] In light of that, and given the widespread sense that pervasive religious diversity really ought to pose a special challenge to religion, we wish to generate a diversity challenge that does not amount to a general recipe for skepticism. We think that such a challenge can be formulated, and that it provides a partial answer to a charge, raised by Peter van Inwagen (2010), that people often hold religious beliefs to stricter evidential standards than other kinds of beliefs, such as political or moral beliefs. Our task will be to explore this challenge.

The explanatory challenge

Like ancient formulations of the problem of evil, the challenge we have in mind starts out as a question:

> **Explanatory Challenge**: Why, if some particular theistic outlook is true, and if it's urgent that everyone believes that outlook, are there so many competing religions in the first place?

In contrast to the epistemological challenges mentioned earlier, the explanatory challenge invites reflection on the origins of diversity. This is of philosophical interest because it turns out that reflection on the origins of diversity is an effective way of uncovering an important tension in the beliefs of many theists—one that has less connection to Smith's questions and more affinities with hiddenness arguments.[3] Our first task will be to say something about what this tension is. Our second task will be to argue that this tension, though already notable in its own right, is harder to resolve in light of recent scientific accounts of religion.[4] We close by considering how critics might respond to our claims. By this point, it should be clear that our goal is not to provide a universal defeater for anyone's theistic beliefs, but to raise what seems to us an important, interesting, and comparatively neglected challenge facing many theists.

The scope of the challenge

Let us begin by clarifying what we are not claiming. We are not claiming that the explanatory challenge arises for everyone. That is to say, our challenge does not rest on the following general principle: for any belief P, if P is controversial and if it is urgent that people affirm P, then there is an epistemic problem in affirming P. This principle is implausible and would only invite the kind of general skepticism that we are seeking to avoid here. After all, take the belief that "humans are largely responsible for climate change" and the belief that "vaccines should be given to children." These beliefs are controversial, at least on the street, and it is urgent that people affirm them, if they are true. Yet, these factors alone do not count against holding them.

Put another way, the explanatory challenge is designed only for those with rather particular background beliefs about God and humanity. Not every person or even every religious person will hold these beliefs. In fact, not every theist will. For instance, those from nontheistic religious traditions, pluralists, nonrealists, or those who are highly apophatic in their orientation, may not face the explanatory challenge; nor will those who conceive of God as highly limited or highly impersonal. In addition, theists who think that God values diversity of opinion more than truth,[5] or simply does not care about what people believe, or does not really want everyone to experience salvation (in the sense of deep, and normally everlasting, union with God) will not face the problem. On the other hand, many theists, we are tempted to say the majority of the world's theists, will reject all such claims as unorthodox and will face the challenge.

For instance, many theists are **truth exclusivists** who affirm that there is a single, uniquely valuable, theistic tradition in the world that by far best approximates the truth about the divine and humanity's place in the world. Some of these theists are further **doxastic exclusivists** who think that believing the right things about God and religion is at least very typically required for salvation.[6] The explanatory challenge is intended for theists who hold at least one form of exclusivism and who conceive God to be a perfect being, with no limitations in knowledge, power, or love. It is further intended for those who understand God's love to mean that God wants everyone to flourish, religiously speaking, not just in the hereafter, but also in the present. For such persons, although it may be crucial to value all people and cultures, and to live together in peace, it will be difficult to equally value all ideas about the divine.[7]

Unsurprisingly, doxastic exclusivists will face the most difficult explanatory challenge, given their claims about the eternal fate of nonbelievers. But it is less often appreciated that even those truth exclusivists who reject doxastic exclusivism as narrow-minded still face a problem here. One reason for this concerns what is truly valuable in a theistic world. As advocates of so-called hiddenness arguments have long pointed out, if a personal God exists, then an explicit divine–human relationship in this life is highly valuable in its own right, even if no one's future salvation is on the line (Schellenberg 1993). As such, it seems plausible to think that a loving God would seek out such a relationship with everyone during their lives and would only permit resistant forms of nonbelief or disbelief in God.

This claim about the value of generic theistic belief is only half the story, however. Although standard hiddenness arguments overlook the point, many theists claim that their particular religious tradition has many unique benefits to offer the world, doxastic and otherwise. For instance, Roman Catholics are increasingly likely to acknowledge that those from other faith traditions, and perhaps even atheists, can experience salvation and can have implicit faith. They will sometimes even grant that Catholics can learn some things about the divine from people in other traditions. But these same Catholics will very often think that there are serious *all things considered* advantages to being Catholic, which offers the benefits of sacramental life, a unique degree of theological knowledge, and communion with the saints.

It is not just traditional Catholics who think this way. Probably most major theistic traditions (with some possible exceptions)[8] have claimed to offer something uniquely valuable to the world, beginning with the truth about the most important matters in the universe. The question that such persons face is this: why would God permit so much diversity if this excludes so many from the temporal and/or eternal benefits mentioned above? True, perhaps some disagreement, say on more peripheral matters within a tradition, is inevitable. We also must not forget that religious traditions do converge on certain ideas.[9] Even so, why so much disagreement even on the core issues? Why have so many extremely populated, and largely incompatible, religious traditions emerged, including African traditional religions, Buddhism, Chinese traditional religions, Christianity, Hinduism, Islam, Jainism, Judaism, Native American religions, Sikhism, Spiritism, and Zoroastrianism?

The explanatory challenge generates an internal tension

We realize that many theists are willing to give answers to these questions and to offer theological explanations of diversity. For instance, in more conservative

Protestant circles, one often hears religious diversity explained by some combination of sin, the Fall, demonic influence in the world, or if you are in certain Calvinistic circles, the selective benevolence of God.[10] We will revisit some of these answers in due course and explore some challenges they face. In the meantime, we hope to underscore why the explanatory challenge is of philosophical interest. Briefly, the explanatory challenge does not merely seek to get theists to think more seriously about why religious diversity exists, it is also an invitation for them to better see a tension in their own beliefs. These beliefs are:

1. A perfect God exists and wants everyone to enjoy the many temporal and/or eternal benefits, doxastic and otherwise, that accompany explicit participation in the one true theistic religion. God is further willing to go to serious lengths to make this happen.
2. God permits multiple, competing, religious traditions to be extremely successful, and thus permits much, if not most, of the world's population to be deceived and to miss out on the temporal and/or eternal benefits in question.

The claim here is not that 1 and 2 are logically inconsistent. For instance, there may be a possible world in which both 1 and 2 are true and in fact one person, at least, has argued that versions of the following two claims are logically possible (Craig 1989).

3. No one who fails to believe the truth about revelation, on account of being raised in a mistaken religion, would have believed the truth even had they been raised elsewhere. Religious diversity thus disadvantages no one, even if doxastic exclusivism is true.
4. An unsurpassably loving God has good reasons for creating lots of people who God knew would not flourish.

We think that 3 is possible, in the broad logical sense,[11] and although we are not confident about 4, we will grant its possibility for the sake of argument. The relevant point is that logical consistency really does not tell us very much. This is because there can be serious tensions in one's beliefs, even if they fall short of logical inconsistencies or contradictions. These tensions can still remain important and troubling; they can also give one good, even if defeasible, reasons to revise one's beliefs. In fact, the present tension seems to be a textbook illustration of this point. If you affirm 1, you have reason to revise 2 and vice versa. If 1 is more fundamental in your web of belief than 2, which in this case seems reasonable, then this clearly puts pressure on some feature of 2, such as doxastic exclusivism.

Although we will not offer a theory of epistemic tensions or epistemic coherence here, we think the basic idea is easy to grasp when one sees it. For another example of an epistemic tension, consider the following two claims.

5. Jamie is a very good and very loving father.
6. Jamie refuses to talk to his children and will not even think about them.

Again, a logically possible story might be told to make both claims come out true (maybe the children have a condition such that talking to them or even thinking about them causes them severe harm). But, again, pointing this out hardly eliminates the tension.

Have we not overlooked something, however? In particular, might a skeptical theist insist that an omniscient God may well be aware of reasons entirely beyond our, often limited, imaginations for why 1 and 2 sit well together? Perhaps. But it is no less likely that God would have reasons, also entirely beyond our imaginations, for why 1 and 2 cannot hope to be reconciled.[12] We thus do not find that factoring in unknown possibilities eliminates a known tension. In saying this, we do not mean to rule out the possibility of mystery. In fact, we think that granting a tension is consistent with having considerable uncertainty about God's ways and so is consistent with so-called skeptical theism—at least, anyhow, when the latter approach is properly understood (Benton, Hawthorne and Isaacs 2016).

Actually, granting the tension seems to us like a fairly modest and reasonable thing to do. For instance, granting the tension does not require thinking that all theists have a full out defeater for any of their beliefs (where a universal defeater requires all theists to abandon some belief). Perhaps a few theists could even manage to avoid partial defeaters or requirements to hold their beliefs less firmly. Much will depend on specific circumstances of a believer and her background evidence and experiences. So the claim, again, is relatively modest. Even so, the claim is strong enough to be interesting. We think the tension we have raised, for instance, could reasonably lead many believers to suffer partial defeat for one or more of their beliefs. In addition, the tension could make it reasonable to give up doxastic exclusivism entirely, and perhaps more core beliefs, even if this is not absolutely required of them.

We realize that some may wish to deny even these weaker claims. But if one really thinks that 1 and 2 are not even in prima facie tension, and that serious diversity does not put even a little bit of pressure on the exclusivist to revise at least one of her beliefs, we confess that we have nothing more to say to such individuals. For those willing to concede the tension, on the other hand, we invite them to read on.

Scientific accounts of religion

Now that the original explanatory challenge is on the table, we hope to update it by incorporating insights from recent scientific accounts of religion. Here we have in mind cognitive and evolutionary approaches to the study of religion, which are discussed in the general area of Cognitive Science of Religion (CSR). Although CSR has troubled some theists, it is not intrinsically atheistic[13] and, interestingly, has been endorsed by many notable theists, including Justin Barrett and Robert Audi.

So what does CSR tell us about religion? Although there is controversy within the field, most people in CSR seem to endorse a version of the **naturalness of religion thesis**. In very broad outline, this thesis claims that religious ideas are found throughout all known human cultures in no small part because humans are naturally predisposed to form and spread beliefs about supernatural agents. Although no one is born with religious beliefs, the basic idea is that with fairly minimal environmental stimuli, instruction, or effort, children in diverse environments tend to come to see the world in terms of some sort of supernatural agency, meaning, and design. Religion, so understood, is thus not merely a product of culture, as many suppose. It is a product of cognition, culture, and, whether directly or indirectly, evolution.

To be sure, there are different ways to put this claim about "naturalness" (Barrett 2010; McCauley and Cohen 2010). There are also interesting debates about how far a CSR framework can explain nontheistic traditions like Buddhism (Pyysiäinen 2003). Most agree, however, that our cognitive capacities and information processing biases constrain which religious ideas get formed and which become cultural. The more interesting challenge is to spell out these claims and to explain why religion arose to begin with. Here, three broad theoretical frameworks have emerged.

First, there is the *by-product interpretation of religion*, which is the standard model at the present time. According to this view, religion is an evolutionary accident. More accurately, religious ideas stem from domain-specific cognitive capacities that evolved in response to nonreligious adaptive challenges, and that fulfill nonreligious functions, but which generate religious ideas (Atran

2002; Bloom 2007). Some of these capacities include agency detection abilities (which helps us to detect agents in our environments), theory of mind (which helps us to attribute beliefs and desires to perceived agents), and teleological reasoning (Evans 2000; Kelemen 2004; Lombrozo et al. 2007). Second, there is the *evolutionary interpretation of religion*. According to this view, religion exists in no small part because it gave groups a prosocial advantage (Sosis 2009; Schloss et al. 2010). Religion thus has adaptive significance. Finally, according to *the hybrid perspective*, the above two perspectives are not in competition and should be integrated. Perhaps religious belief starts out as a by-product. But once it gets going, it can help small groups to survive and to make the transition to large, cooperative, anonymous societies (Norenzayan 2013). If that is right, then by-products can later assume functions (Powell and Clarke 2012), suggesting an intricate interplay between genes and culture.

How CSR explains religious diversity

Interestingly, no matter which of the above three views one chooses, serious religious diversity seems almost inevitable. To see why, we admittedly need to engage in some storytelling—that is, speculative imaginings about what steps might have taken place in the past. We do not think that this is a bad thing, however. Science needs stories. If these stories lack sufficient evidence or explanatory payoff in the long run, or if they prove fundamentally misguided, we should reject them. Otherwise, we should refine and improve them over time. So what do the stories that are told in this context typically look like? Beginning with the by-product view, one often encounters stories like the following.

Story A: Approximately 40,000 years ago, a certain individual, walking in a forest at night, was suddenly stopped by a strange noise. The experience activated her agency detection capacity and she was suddenly filled with the sense that someone was watching her. Seeing no person or animal around, she quickly formed a belief in what later came to be called invisible forest spirits. She told her group about the experience and her testimony was well received. Her story further activated the theory of mind capacities of those present, leading to various questions such as Who are these spirits? What do they want? Are they our ancestors? Over time, a more refined interpretation of spirits developed.

> **Story B**: Approximately 10,000 years ago, a different person lived in a stormy region near the fields. Following some violent cracks of lightening during a particularly bad storm, he formed the idea that a powerful being ruled the local skies and might be angry with him. He reported his experience back to his community and his testimony was accepted, eventually giving rise to more specific ideas and rituals, including a rain dance. Over time, and with the help of transmission errors, some of his descendants slowly switched from henotheism (the idea that each region has its own high god) to monotheism (the idea that there is but one God that rules the universe).

According to this by-product perspective, our agency detection systems and our theory of mind will bias us toward anthropomorphism (Guthrie 1993). Even so, the specific religious ideas that emerge and get transmitted, for instance the name of a spirit or what it wants, are highly context-dependent (Schloss et al. 2010, 626), ensuring a certain amount of diversity.

What about the evolutionary view of religion? Once again, nothing about this view predicts that different groups will have the *same* detailed religious beliefs or practices. Recall the core evolutionary idea. A belief in invisible, monitoring, agents helps to promote prosocial behavior in a group, which, in turn, creates a survival advantage. Since these supernatural watchers could have different identities (i.e., they could be witches, gods, or ancestor spirits etc.), and yet presumably play a similar functional role, we have no reason to expect only a belief in one kind of watcher to emerge. In fact, recent experimental evidence suggests that different supernatural agents can effectively play the watching role. In one influential study, being primed with ideas about Princess Alice was sufficient to inhibit cheating behavior in children (Piazza et al. 2011). As for the hybrid view, we do not see how this alternative would change the basic story we have been telling. It is still going to be true that people's religious ideas will seriously differ. In a word, if neither the by-product view nor the evolutionary view gives us reason to expect strong agreement on the details, combining them will not either.

The updated explanatory challenge

If the above claims about CSR are on track, then we can update the explanatory challenge as follows:

> **Updated Explanatory Challenge**: Why, if some particular theistic outlook is true, and if it's urgent that everyone believes that outlook, would God design the world in such a way that serious religious diversity naturally arises?

Whereas the original explanatory challenge asks why God would tolerate or permit so much diversity, the updated challenge asks why God would explicitly design the world to include it. This difference matters to the tension we have been exploring. Indeed, we shall now offer three reasons why the updated challenge, which makes use of CSR, is harder to answer than the original version.

> **Reason One**: The updated explanatory challenge puts new pressure on many Abrahamic theists to give up what is perhaps the most commonly cited theological explanation of religious diversity, namely human free will and moral shortcomings.

Although not all theists are comfortable with it, many Abrahamic theists attribute religious diversity to sin and free will. This view, often held by ordinary believers, is also alive and kicking in much analytic philosophy of religion. Consider the following passage from Alvin Plantinga,

> Were it not for sin and its effects, God's presence and glory would be as obvious and uncontroversial to us all as the presence of other minds, physical objects, and the past. Like any cognitive process, however, the *sensus divinitatis* can malfunction; as a result of sin, it has indeed been damaged. (2000, 177)

To be fair, Plantinga does not assert that this view is true, only that it is true if his model of religious knowledge is true. Even so, he clearly believes it and notes that he is in good theological company. A similar view is held by some cognitive scientists. For instance, Justin Barrett makes an appeal to the Fall to answer the following very good question:

> If … God created humans so that they might enjoy a relationship with Him, why would God leave such important cognitive capacities to chance plus natural selection … why do the documented conceptual biases only encourage belief in superhuman agents generally and not in one true, accurate god concept? … why not hard-wire into our brains a fully formed belief in God? (2009, 97)

Although Barrett does not develop this challenge, he briefly entertains the idea that somehow (we are never told exactly how), despite CSR and evolution, the "diversity of god concepts we see is a consequence of human error and not divine design" (2009, 97). Barrett is clearly referencing the Fall here and notes that a "sinful, fallen world" corrupts our concepts.[14] This claim, like Plantinga's, does not sit well with CSR. In fact, it appears inconsistent with Barrett's own claim, made elsewhere, that "arguably the oldest and most widespread form of god concepts is the ancestor spirit or ghost, a type of afterlife belief" (2007, 775). On the latter view, belief in limited spirits and gods is the default state, with a belief in big gods coming later. Since all of this looks perfectly natural, CSR provides evidence for the following counterfactual: even if humans never behaved badly, there would still be pervasive religious diversity.

To clarify, in raising the above worry, we are not claiming that no diversity results from sin, given theism. For example, Augustine seems to have thought that God caused diversity at certain points to confuse the proud (De Doctrina Christiana II, 1–15). Perhaps the theist could call on this, and other kinds of spiritual resistance, to account for some religious diversity. But one would be hard pressed to explain all or most religious diversity this way. For one thing, there is the sheer amount of diversity that needs explaining: for any particular religious outlook, most of the world's population does not subscribe to it. For another thing, the typical causes of diversity look to be largely outside the scope of human agency. We do not typically choose our cultures, environments, beliefs, or our cognitive processes. In addition, there is the explanatory worry posed by an uneven distribution of religious ideas. The world's cultures are comparably prideful or sinful, and yet, the demographics of theism vary substantively over place (Maitzen 2006) and time (Marsh 2013), with early humans apparently lacking any theistic concepts. For the doxastic exclusivist, who thinks that one religion is uniquely true, the demographics of the target outlook become even more lopsided.

Some of the above considerations about the Fall and the demographics of theism have been noted by others, including Stephen Maitzen. Since Maitzen and others tend to entirely neglect the question of CSR and the origins of religion, however, they overlook cognitive and evolutionary explanations of the demographics of religious belief. This makes a difference. For Maitzen's challenge, aside from lacking a developed naturalistic explanation of the demographics of theism,[15] leaves open whether God merely allows an uneven distribution of belief or causes it. By contrast, the updated explanatory challenge takes a clear stand on this matter. Indeed, the second reason why this updated challenge is more significant than more traditional versions draws upon the distinction between doing and allowing.

> **Reason Two**: The updated explanatory challenge suggests that God, given theism, does not merely allow serious religious diversity to emerge, but, to a large extent, causes it. This is because if theism is true, God is behind nature and responsible for creating brains and environments like ours through natural processes. But then serious religious diversity is largely the product of divine design.

Many theists think there is a moral difference between doing and allowing and will not be well positioned to reject reason two. Evidence for this claim may be found by looking at discussions of the problem of evil. It is widely thought by philosophers, including theists, that the problem of natural evil is harder to answer than the traditional versions. There are two reasons for this. First, the main classical theistic explanation of evil (free will) is no longer available. Second, it can be difficult not to attribute natural evil to divine action. We think CSR does for the explanatory challenge of religious diversity what natural evil does for the traditional challenge from evil. This is not to deny that there are ways of thinking about God, causation, and providence that dissolve the distinction between doing and allowing—for instance, occasionalism. But most theists seem to reject such views in part because they resist the idea that a perfect being would be so causally connected to evil actions.

To be clear, in raising this CSR-based worry that religious diversity is a natural feature of the world, we are not claiming our brains wholly determine our religious concepts, given CSR. As Michael Murray notes,

> .... the mental tools identified by cognitive scientists [do not] give rise to these inconsistent [religious] beliefs *all on their own*. When environmental conditions stimulate HADD, Theory of Mind, memorability and transmissibility via minimal counterintuitiveness, etc., the outputs of these tools are still highly non-specific. HADD tells me there is "an agent"; my beliefs about what sorts of fauna inhabit these parts lead me to conclude that the agent is a bear or a tiger or the bogeyman .... no one doubts that divergent cultural traditions play an enormous role in giving religious concepts their specific contours. (2009, 172)

We agree that environment matters to religious belief. In fact, it is the interplay between cognition and culture that, interestingly, poses much of the problem. For God, given theism, gets to decide how much of a role environment plays in the formation of religious belief and gets to decide how robust religious cognition is. To see the point, consider three possible worlds. In the first world, God hardwires some fairly detailed religious ideas into people's minds, such that people are practically born believing. In the second world, God does not make specific religious ideas innate, but ensures that every person with an open mind and an open heart acquires a detailed grasp of the one true religious tradition through experience. In the third world, God gives people general dispositions to form supernatural beliefs, but mostly leaves the details up to cultural evolution and environmental differences.

Now one could debate which of the first two worlds is preferable, given theistic exclusivism. For instance, some may wish to argue that world two is the preferable world because of the value of discovering important truths gradually. Others may prefer world two because they wonder whether universal agreement, owing to innate cognitive structure, would make some anxious about the possibility of universal deception or debunking. By contrast, others might insist that when it comes to the most urgent matters of the universe, at least (like certain moral ideas or a tendency to recognize one's parents at birth) that innate knowledge could be rather helpful. Failure to see this in the case of theism, goes the thought, is due to a status quo bias.

We are fully prepared to acknowledge that there could be a reasonable debate about the relative merits of worlds 1 and 2. What seems less plausible to us is the suggestion that world 3 (i.e., the world we actually find ourselves in given theism and CSR) is preferable to both of these other worlds. For notice that the third world, under exclusivist assumptions, has an important disadvantage that the first two worlds lack. If God actualizes world three, substantive religious diversity, and by extension religious ambiguity, will naturally arise as a consequence of God's creative choice. In the case of doxastic exclusivism, many will miss out on salvation. In the case of truth exclusivism, many will walk in darkness during their time on earth.

The third and final sense in which CSR has normative or philosophical significance goes as follows:

> **Reason Three**: Once we appreciate that some of the main competitors to theistic religion, namely naturalism, deism, and pluralism, do a better job of explaining the origins of religious diversity than Abrahamic theism, this might be thought to lend comparative support to these other views.

Reason three does not claim that Abrahamic religion, as opposed to say generic theism, makes the bare fact of religious diversity comparatively surprising. After all, Jewish, Christian, and Islamic scriptures are full of stories about religious differences—in fact these scriptures further acknowledge the reality of an uneven distribution of theistic belief. But then these traditions entail that there is lots of religious diversity (rendering the probability of such diversity 1).

Given this way of reasoning about probability (Otte 2012), the bare existence of religious diversity, and even, pace Maitzen, the uneven distribution of theistic belief, cannot generate an empirical problem for Abrahamic faiths. For the reality of such diversity, if anything, confirms these views.

It is important to see that the above kind of reasoning, even if endorsed, does nothing to undermine reason three. For reason three is not about the bare fact of religious diversity, or its distribution, but about how diversity arises in the way that it does. Nothing in Abrahamic scriptures implies that religious diversity arises naturally in the way CSR suggests and on the time-scale it suggests. In fact, many theologians will insist, to the contrary, that religious diversity and the confusion it creates is unnatural. Many Christians and Jews, for instance, have thought that unbelief in a higher God is always foolish, sinful, and without excuse, given natural revelation.[16] We appreciate that others will resist these interpretations. We also appreciate that exegesis is no easy task, and that many work hard to show that scripture sits well with science. Even so, the bare fact that there is a controversy within these major theistic traditions is significant. Since there is no similar controversy arising from naturalism, deism, or pluralism,[17] this fact seems to support these other views.

Objections

We will now consider some objections to our arguments. We shall restrict our focus to just three.

> **Objection 1**: Even if all of the scientific claims you discuss are true, which some might doubt, you still have not shown that religious diversity largely results from divine design. After all, in response to worries about evolution and natural evil, Alvin Plantinga has toyed with the idea that supernatural agents, besides God, could be responsible. He states that "Satan and his minions" might have been "permitted to play a role in the evolution of life on earth, steering it in the direction of predation, waste, and pain (2011, 59)." Perhaps this idea could also be invoked to explain religious diversity. Here religious diversity will be natural all right, but it will be due to demonic agency and not to God.

So what should we make of the idea that demons have this kind of power over creation? Of course, the suggestion will seem absurd to those who do not believe in demons to begin with, as Planting appreciates, but that is not the right standard to adopt when considering an internal challenge. Even when we look at the matter internally, however, problems arise. Perhaps most notably, Plantinga's suggestion is difficult to square with the traditional idea that creation starts out good and indeed very good—on Plantinga's demonic evolutionary hypothesis that does not seem to be true.[18] In addition, that so many of Plantinga's theistic colleagues would reject his claim might give him further reason to question it.

A second possible objection, based on the work of Peter van Inwagen, also defends the idea of serious discontinuity in natural history and specifically human evolution.

> **Objection 2**: The updated explanatory challenge, at least, fails since it remains epistemically possible that early humans were full-blown monotheists and that diversity arose because of sin. This is because it is epistemically possible, or true for all anyone knows, that the following story from van Inwagen is true: "for millions of years perhaps for thousands of millions of years, God guided the course of evolution so as eventually to produce certain very clever primates, the immediate predecessors of *Homo sapiens*. At some time in the last few hundred thousand years, the whole population of our pre-human ancestors formed a small breeding community—a few thousand or a few hundred or even a few score .... In the fullness of time, God took the members of this breeding group and miraculously raised them to rationality. That is, he gave them the gifts of language, abstract thought, and disinterested love—and, of course, the gift of free will ... God ... also took them into a kind of mystical union with himself, the sort of union that Christians hope for in Heaven and call the Beatific Vision. But, somehow, in some way that must be mysterious to us, they were not content with this paradisal state. They abused the gift of free will and separated themselves from their union with God" (2006, 85–86). All of this, one might add, led to confusion about who God is and, for unknown reasons, the confusion became more pronounced in some regions than others.

There is much one could say here. Many biologists will claim population genetics, or other areas of science, rule out such possibilities. Many philosophers will think that the prior probability of van Inwagen's story is extremely low, even for ideal agnostics, and that it has more in common with skeptical scenarios than with a serious proposal. We will let van Inwagen fight these battles with others. A more basic problem with the current objection is that it still leaves the tension we raised basically intact. After all, van Inwagen would presumably admit that it is also epistemically possible—even for many theists—that the above story is false, and that diversity is indeed natural. But that alone is sufficient to establish the tension. Consider an analogy: knowing that it is epistemically possible that one does not have cancer is preferable to lacking this knowledge, to be sure.

But one would surely still have reason to worry if the hypothesis that one had cancer became epistemically possible. In that case, one would want compelling evidence that one is cancer-free. We think something similar applies to our challenges. What goes under the name of a "defense" in discussions about the problem of evil is not enough to answer them.

There are naturally other objections that could be raised against our claims. But we shall take up only one more.

> **Objection 3**: In focusing on Plantinga style theologies, your discussion leaves out other, often more positive, conceptions of the Fall and religious diversity. For instance, according to Irenaeus, humanity did not begin in a state of perfection, but in a less than fully mature state. As for humans, they gradually lost their innocence, but never were perfect to begin with. In fact, on this view, all of creation needs to gradually get better, which sits fine with CSR. Your focus on sin as a theological explanation of diversity, in addition, is not the whole story. For example, Augustine suggests that diversity results from (in addition to sin) the sheer fact of human reality, where things as such are not easily graspable. There are also Islamic interpretations of diversity, according to which monotheistic diversity exists, in part, to motivate healthy moral competition among different monotheistic groups. Finally, maybe religious diversity exists, in part, because it helps people to engage with truth—which is basically the rabbinical model of truth in contestation.

We agree that there are many different possible religious explanations of diversity and wish we could spend more time canvasing different views. That said, we do not see a clear objection here. We never claimed that Plantinga's conception of the Fall and diversity is the only conception, only that it represents the dominant conception among philosophers of religion and more traditional theologians, particularly in the West. Showing that the dominant conception risks being disconfirmed seems significant to us. We suspect that the next most popular explanation of religious diversity, historically, is the Tower of Babel story (Seely 2001, 19).[19] This explanation too is challenged by CSR. But we resisted the temptation to explore this, since fewer people read the story of Babel literally nowadays.

As for Irenaeus's views of creation and Fall, it is true that this may be compatible with CSR than more common views (De Cruz and De Smedt 2013). But it only restates the normative problem. After all, it is tempting to think that if diversity arises, under exclusivist assumptions, it *should* be owing to clear human moral shortcomings and not to divine design.[20] The idea that God would make the world such that diversity naturally arises even though it is extremely bad for people's flourishing, which exclusivist views seem committed to, recall, *just is* our second tension. It is precisely because Irenaeus' view is more scientifically friendly, in other words, that it succumbs to our updated explanatory challenge. Of course, one could always revise one's theology in a pluralist direction to escape the tension, as John Hick seems to have done in his later years. But that would only be to concede the force of the explanatory challenge.

As for Augustine's idea that reality is not easily graspable by limited human beings, there is something to this. We also like the idea, shared by some Muslim scholars, that diversity can motivate healthy moral competition among monotheistic groups (Shah-Kazemi 92)[21]—and further find the rabbinical idea of truth in contestation interesting.[22] It should be pretty clear that none of these answers gets us out of the pickle, however. Beginning with Augustine's claim about how hard it is to grasp reality, this claim leaves too much unexplained. Why is not everyone comparably in the dark about the divine, given human limitations? Why are some persons, and indeed some entire groups, so much more sensitive to the truth than others? And why does geography have so much to do with it, if a providential God exists and extends grace to all? Calling on our epistemic limitations does not explain these patterns. By contrast, many of these same patterns are to be expected given CSR.

As for the idea that diversity stirs up healthy moral competition, it is not obvious that monotheistic groups are outperforming other groups, morally—at least not by the kind of margin that would be required for this view to work. Besides, the view leaves wholly unexplained why nonmonotheistic religions exist. It will not work to expand the view so it becomes an explanation of all religious diversity either, since religious groups are not obviously more virtuous than secular groups: although they do better in some areas, like giving to charity, they do worse in others. The present view also raises a difficult question. Why would God design the world such that two very valuable outcomes—a detailed acceptance of revealed truth and moral motivation—are in competition in the first place?

The idea that there is truth in contestation too runs into empirical worries. For one thing, it can seem that most people do not contest the truth very much. Most seem rather content with their own outlooks and would literally rather not open the door to someone from a different tradition. As for the sizable minority who do enjoy a good religious discussion, these rarely convert to another tradition in response to their contestations (as is widely known, conversion across world religions is generally rare). All of this makes it hard to say that religious diversity exists, despite truth exclusivism, *because* it makes people more seriously engage with the truth or *because* it makes people come into contact with more religious truth in the long run. There are naturally other Jewish perspectives on diversity, including the idea that God, in some sense, "*wanted* Christianity and Islam to arise and spread, even if his reasons … are hidden" (Shatz 2011, 372). But this, again, does not explain nonmonotheistic diversity; it further leaves monotheistic diversity rather mysterious.

In short, all the theological explanations of diversity of which we are aware either risk being falsified by CSR or other empirical factors, or do not actually help resolve the normative theological tension of why there should be so much diversity, again, given either form of exclusivism. We thus invite more discussion on these matters.

Conclusion

In conclusion, we have accomplished two things in this chapter. First, we explored a diversity challenge that does not create general skeptical problems for nonreligious beliefs. Second, we showed that CSR intensifies this challenge and so has philosophical significance for this reason. This last point about CSR is worth stressing because, although we did not mention it above, it is not just standard diversity challenges that threaten to generalize. Many philosophers have thought that standard debunking challenges to religious belief, based in CSR, also generalize. We were able to resist these two possible bases for skepticism.

Now some might have thought that CSR is, in reality, good news for theism. After all, this science seems to sit well with something that many theists will affirm: namely that religion is not *merely* a contingent cultural phenomenon, but a deep feature of human nature—and therefore something that a loving God could get behind. This tempting suggestion, popular among some evolutionary theists, ignores a more specific and more accurate way of describing what is going on, however.

Our explanatory challenge was not directed at supernaturalism in general, nor was it directed at every possible version of theism. Rather, it was directed at the actual beliefs of many theists when they say or imply that religious diversity is a highly unfortunate feature of the world, or when they claim that religious diversity is valuable but remain truth exclusivists. Although diversity no doubt makes the world a far more interesting place in many respects, relative to the particular beliefs we discussed, its existence is puzzling. If what is really part of human nature turns out to be a vague sense of the supernatural, a sense that can easily be taken in various directions, this more specific fact might be thought to be bad news for theists who are doxastic exclusivists, truth exclusivists, or both. If these claims are on track, then although some theists might find ways to preserve the rationality of their beliefs, despite the challenge we have raised (and recall we did not deny this), it is hard to say that CSR clearly has no prima facie philosophical significance for popular versions of theism or for claims commonly made by theists.

Finally, we should acknowledge that CSR, like any young science, could end up being completely wrongheaded. But our task here was not to defend CSR, only to discuss its implications on the assumption that its core claims are correct. To us, CSR's core claims look promising (much more so than traditional theological accounts of diversity) but still need further confirmation. Whatever happens with CSR, though, our stripped down explanatory challenge remains pressing and, in comparison to Joseph Smith's questions at least, neglected. It surely warrants future exploration.[23]

Notes

1 Smith eventually prayed for wisdom and claimed to discover the truth in a rather spectacular vision. Unfortunately, most who employ his method of belief formation are not so lucky and often come away with a very different outlook.

2 For instance, our moral and political beliefs are often contingent (we would believe them to be false had we been born elsewhere), controversial (many apparent peers reject them), and probabilistically challenged (the competition seems to stack against them). Now maybe the lesson here is that lots of our beliefs are in trouble on account of failing these various tests. But few epistemologists are willing to draw this more global lesson. Most prefer to see a problem with the diversity challenges.

3 We do not think the challenge here is the challenge from divine hiddenness, however, in part because nonbelief could arise, in principle, even if there was no diversity of religious opinion (for instance, everyone might be atheists or nontheistic Buddhists). Similarly, hiddenness arguments tend to focus on the value of belief in generic theism, as opposed to more specific beliefs and practices, and tend to say almost nothing about the origins of religion. We also do not think our challenge is reducible to the so-called soteriological problem of evil, since our challenge would go through even if all persons are saved. In any case, perhaps our challenge is best seen as a hybrid between a diversity challenge and a hiddenness argument, albeit one that introduces new features, as we shall see. In the end, we do not care too much about labels, however.

4 At least, that is, for those who find these accounts plausible. We cannot defend entire scientific approaches here.

5 It is not enough to think that God values diversity to escape the problem; one must also show that diversity itself, or some good that diversity makes possible, is more valuable than true belief and what it makes possible.

6 Doxastic exclusivists will often grant that there is far more to religion than belief but will add that belief remains extremely important. Those who think good moral behavior is sufficient for salvation are not doxastic exclusivists, though they may still be truth exclusivists.

7 William Lane Craig, a well-known doxastic exclusivist, thus states: "compassion toward those in other world religions is therefore expressed, not in pretending that they are not lost… but by… making every effort… to communicate to them the life-giving message of salvation" (1989, 187–188).

8 This explains the many missionary and proselytizing efforts throughout history. There is the idea of Da'wah in Islam and Christian evangelism. Some even argue that Hinduism can, despite initial appearances to the contrary, be seen as a missionary religion (Sharma 2011). Judaism is perhaps the most obvious counterexample here—and there are many historical, sociological, and religious reasons for this (Shatz 2011). Even so, if some traditional strand of Judaism were literally true, many non-Jews might have an interest in knowing this.

9 For example, some religions converge on monotheism, and many more converge on the golden rule.

10 We will say nothing about Calvinistic views in this chapter mainly because we are talking to theists who think God loves everyone and fully desires their salvation.

11 One of us once argued that a version of claim 3 could be seem minimally reasonable for Molinist theists who are also universalists about salvation (Marsh 2008). We do not endorse 3 as serious possibility, in part, because we do not like what it seems to imply about many nontheists. In any case, Craig's Molinist explanation of 3, in rejecting universal salvation, raises even bigger worries.

12 This kind of point is often made by Paul Draper.

13 Some might worry that CSR is methodologically atheistic and claim that it simply begs the question against the theist for this reason. After all, it might be thought that the naturalistic approaches to religion leave no room for some groups to have received special miracles, moral instructions, and Holy Books from on high. In response: this objection, though understandable given how some scientists speak, is unfortunately misguided. For whatever particular scientists do or say, CSR can in principle allow that some miracles or revelation might have occurred, if the evidence favors this. The problem is that, even given this openness to the supernatural, much diversity remains. By the theist's own admission, then, miracles, revelation, and the testimony of the faithful have not been especially

effective in eliminating diversity. But then pointing to methodological worries about science is a moot point.

14 He even adds that in the absence of human moral failure, "perhaps children would inevitably form a perfect concept of God" (2009, 97).

15 Maitzen notes briefly on two occasions that, on naturalistic explanations, "culture and politics alone" (2006, 185) determine and explain the demographics of theism. Though Maitzen's overall challenge is interesting, this strikes us as too simple.

16 Psalm 14 and Romans 1 have often been read this way, for example.

17 We have heard some reject this claim with the following counterfactual: if a perfect God did not exist, nothing would exist. We will resist the temptation to explain why we think this counterfactual, even if taken to be true, does not resolve our challenge.

18 We are indebted to John Greco for helping us to see this point.

19 Many Jewish and Christian thinkers, including Augustine, Luther, and Calvin, apparently understood the Tower of Babel to explain much cultural and linguistic diversity (Seely 2001, 19). We have heard some religious leaders invoke this same explanation to account for religious diversity.

20 This is not to say that "free will views" face no challenges, only that they face fewer challenges than divine causation views.

21 We are told that Qur'an 5:48 is commonly cited as evidence for this view.

22 Thanks to Sarah Coakley for drawing our attention to this rabbinical view.

23 We are grateful to Sarah Coakley, Helen De Cruz, John Greco, Jamie Schillinger, Corliss Swain, Charles Taliaferro, and to two anonymous referees for their feedback and questions. The ideas discussed here are ours and should not be assumed to reflect the views of any institutions with which we are affiliated.

Bibliography

Atran, S. 2002. *In Gods We Trust. The Evolutionary Landscape of Religion*. Oxford: Oxford University Press.

Augustine of Hippo. 1995. *Augustine De doctrina Christiana*. Oxford: Clarendon.

Barrett, J. L. 2007. Cognitive science of religion: What is it and why is it? *Religious Compass* 1(6): 768–786.

Barrett, J. L. 2009. Cognitive science, religion and theology. In M. Murray and J. Schloss, eds, *The believing primate: Scientific, philosophical, and theological reflections on the origin of religion*, 76–99. New York: Oxford University Press.

Barrett, J. L. 2010. The relative unnaturalness of atheism: On why Geertz and Markusson are both right and wrong. *Religion* 40: 169–172.

Benton, M. A., J. Hawthorne, and Y. Isaacs. 2016. Evil and evidence. In J. Kvanvig, eds, *Oxford studies in philosophy of religion*, vol. 7: 1–31. Oxford: Oxford University Press.

Bloom, P. 2007. Religion is natural. *Developmental Science* 10: 147–151.
Bogardus, T. 2013. The problem of contingency for religious beliefs. *Faith and Philosophy* 30(4): 371–392.
Craig, L. W. 1989. "No other name": A middle knowledge perspective on the exclusivity of salvation through Christ. *Faith and Philosophy* 6: 172–188.
De Cruz, H. and J. De Smedt. 2013. Reformed and evolutionary epistemology and the noetic effects of sin. *International Journal for Philosophy of Religion* 74: 49–66.
Evans, E. M. 2000. Beyond scopes: Why creationism is here to stay. In P. L. Harris, C. N. Johnson, and K. S. Rosengren, eds, *Imagining the impossible: Magical scientific, and religious thinking in children*, 305–331. Cambridge: Cambridge University Press.
Feldman, R. 2007. Reasonable religious disagreements. In L. M. Antony, ed., *Philosophers without Gods: Meditations on Atheism and the Secular Life*, 194–214. New York: Oxford University Press.
Guthrie, S. 1993. *Faces in the clouds: A new theory of religion*. New York: Oxford University Press.
Hick, J. 1997. The epistemological challenge of religious pluralism. *Faith and Philosophy* 14: 277–286.
Kelemen, D. 2004. Are children "intuitive theists"?: Reasoning about purpose and design in nature. *Psychological Science* 15: 295–301.
Kelly, T. 2005. The epistemic significance of disagreement. In T. Gendler-Szabo and J. Hawthorne, eds, *Oxford studies in epistemology*, 167–196, Vol. 1. Oxford: Oxford University Press.
Lombrozo, T., D. Kelemen, and D. Zaitchik. 2007. Inferring design: Evidence of a preference for teleological explanations in patients with Alzheimer's disease. *Psychological Science* 11: 999–1006.
Maitzen, S. 2006. Divine hiddenness and the demographics of theism. *Religious Studies* 42: 177–191.
Marsh, J. 2008. Do the demographics of theistic belief disconfirm theism? A reply to Maitzen. *Religious* Studies 44: 465–471
Marsh, J. 2013. Darwin and the problem of natural nonbelief. *The Monist* 96: 349–376.
McCauley, R. N. and E. Cohen. 2010. Cognitive science and the naturalness of religion. *Philosophy Compass* 5: 779–792.
Murray, M. 2009. Scientific explanations of religion and the justification of religious belief. In M. Murray and J. Schloss, eds, *The believing primate: Scientific, philosophical, and theological reflections on the origin of religion*, 168–178. New York: Oxford University Press.
Norenzayan, A. 2013. *Big gods: How religion transformed cooperation and conflict*. Princeton, NJ: Princeton University Press.
Otte, Richard. 2012. Comparative confirmation and the problem of evil. In J. Chandler and V. S. Harrison, eds, *Probability in the philosophy of religion*, 127–143. Oxford: Oxford University Press.

Piazza, J., J. Bering, and G. Ingram. 2011. "Princess Alice is watching you": Children's belief in an invisible person inhibits cheating. *Journal of Experimental Child Psychology* 109: 311–320.

Plantinga, A. 2000. *Warranted christian belief.* Oxford: Oxford University Press.

Plantinga, A. 2011. *Where the conflict really lies?: Science, religion, and naturalism.* New York: Oxford University Press.

Powell, R. and S. Clarke. 2012. Religion as an evolutionary byproduct: A critique of the standard model. *British Journal for Philosophy of Science* 63: 457–486.

Pyysiäinen, I. 2003. Buddhism, religion, and the concept of "God". *Numen* 50(2): 147–171.

Schellenberg, J. L. 1993. *Divine hiddenness and human reason.* Ithaca, NY: Cornell University Press.

Schellenberg, J. L. 1997. Pluralism and probability. *Religious Studies* 33: 143–159.

Schloss, J., J. Barrett, and M. Murray. 2010. Looking past vs. overlooking cognitive-evolutionary accounts of religion: A response to Nathaniel Barrett. *Journal of the American Academy of Religion* 78(3): 622–628.

Seely, P. H. 2001. The date of the tower of Babel and some theological implications. *Westminster Theological Journal* 63: 15–38.

Shah-Kazemi, R. 2013. Beyond polemics and pluralism. The universal message of the Qur'an. In M. Hassan Khalil, eds, *Between heaven and hell: Islam, salvation, and the fate of others*, 87–105. New York: Oxford University Press.

Sharma, A. 2011. *Hinduism as a missionary religion.* Albany, NY: State University of New York Press.

Shatz, D. 2011. A Jewish perspective. In C. V. Meister, ed., *The Oxford handbook of religious diversity*, 365–380. New York: Oxford University Press.

Smith, J. 1902. *History of Joseph Smith: The Prophet.* Salt Lake City: Deseret News.

Sosis, R. 2009. The adaptationist-byproduct debate on the evolution of religion: Five misunderstandings of the adaptationist program. *Journal of Cognition and Culture* 9: 315–332.

van Inwagen, P. 2006. *The problem of evil.* Oxford: Oxford University Press.

van Inwagen, P. 2010. We're right. They're wrong. In R. Feldman and Ted. A. Warfield, eds, *Disagreement*, 10–28. Oxford: Oxford University Press.

5

A Cognitive Psychological Account of Reasoning about Ritual Efficacy

Cristine H. Legare, Rachel E. Watson-Jones, and Andre L. Souza

Introduction

Ritual is a pervasive feature of human behavior, yet has been understudied from a psychological perspective. Until recently (Atkinson and Whitehouse 2011; Legare and Souza 2012, 2014), ritual has been studied primarily by anthropologists using qualitative methodologies, making it difficult to establish robust generalizations about the causes and effects of features of rituals on social cognition and behavior (Rappaport 1999; Rossano 2012).

The complexity and diversity of rituals is bound by both cognitive and socioecological constraints (Atkinson and Whitehouse 2011). For example, different elements of ritual form have been associated with costly signaling (Irons 1996; Sosis and Alcorta 2003; Sosis et al. 2007), obsessive compulsive disorder and the human hazard precaution system (Boyer and Lienard 2006), cognitive constraints on memory systems (Whitehouse 1995), the role ascribed to supernatural agency (McCauley and Lawson 2002), modes of codification and transmission (Whitehouse 1995), and the scale and structure of religious communities (Gellner 1969; Goody 1986; Werbner 1989; Whitehouse 2001).

The development of ritual studies as an independent and interdisciplinary area of scholarly study demonstrates the complexity of ritual as an analytical tool and as a universal human experience. Theoretical discourse on ritual has also focused primarily on belief and action, in isolation and interaction (Bell 1992). Rituals often represent sacred beliefs, express inner states of feeling and emotion, symbolize theological ideas or social relations, and invoke psychophysical states (Shore 1996; Csordas 2002; Ruffle and Sosis 2003; Sax 2010; Sax et al. 2010; Whitehouse and Lanman 2014). Whitehouse and Lanman (2014) argue for the "fractionating strategy" for studying ritual. They propose that, "Under this folk category [ritual] are numerous cognitively and behaviorally universal patterns that are normally associated with the term 'ritual,' including such phenomena as synchronic movement, causally opaque action, and both euphoric and dysphoric arousal" (675). Like similar trends in the study of religion (Lawson and McCauley 1993), the fractionating strategy opens the door for a scientific account of how various phenomena normally associated with the folk term "ritual" influence thought and behavior.

Across cultures, and the historical record, rituals are widely used for protective, restorative, and instrumental purposes (Sørensen 2007; Sax et al. 2010). Records of rituals used for problem-solving purposes date back to ancient Egypt (*The Papyrus Ebers* 1931, 1937), and the use of rituals to treat problems as diverse in etiology as asthma and unemployment is widespread in contemporary cultural contexts such as the United Kingdom (Hutton 1999), the United States (Crowley 1989), Brazil (Cohen and Barrett 2008; Souza and Legare 2011; Legare and Souza 2012, 2014), and South Africa (Ashforth 2001; Legare and Gelman 2008). If rituals are used to solve problems, the individuals engaging in ritual must consider the causal efficacy of the actions involved (Csordas 2002; Sax 2004; Sax et al. 2010). Ritual, however, is opaque from a physical-casual perspective. How do individuals evaluate the efficacy of ritual actions in the absence of direct causal information into their effectiveness?

The objectives of this chapter are threefold. First, we provide a cognitive psychological account of the use of ritual for instrumental, problem-solving purposes. Next, we provide a review of new psychological research on reasoning about the efficacy of ritual. Finally, we discuss evidence for the relationship between ritual and perceptions of control.

The use of ritual for problem-solving purposes: Reasoning about ritual efficacy

Rituals, which we define as socially stipulated, causally opaque behavior (Legare and Herrmann 2013), present a challenge to theoretical accounts of causal reasoning because they are not defined in terms of physical-causal mechanisms or processes. Even when rituals are widely understood in the context of a certain belief, there is often not an expectation of a direct causal connection between the set of actions involved in the ritual and the material outcomes that are desired and observed (Sørensen 2007; Schjoedt et al. 2013). Because rituals are not bound by the same kinds of intuitive physical-causal constraints that characterize non-ritualistic actions, they can be considered *irretrievably causally opaque*. Thus, even though rituals are intended to have particular effects on the world (e.g., rituals used to promote crop fertility or to heal disease), they are not expected to do so by means that are transparent or even in principle knowable based on principles of physical causality.

Even though rituals cannot be interpreted from the perspective of physical-causal reasoning, rituals used for problem-solving purposes still reflect intuitive beliefs about causal reasoning and the efficacy of goal-directed action sequences. Consider Tambiah's (1979) classic definition of ritual—"patterned and ordered sequences of words and acts, (…) characterized in varying degrees by formality (conventionality), stereotypy (rigidity), and redundancy (repetition)." In previous work, we have proposed that the defining characteristics of ritual are the product of an evolved cognitive system of intuitive causal principles (i.e., that repetition, number of procedural steps, and the specificity of procedural detail will lead to efficacious outcomes) (Legare and Souza 2014). In the section that follows, we provide a review of recent experimental evidence supporting this proposal.

The historical and ethnographic record has documented substantial variation in the content, practices, and artifacts used in rituals around the world and over historical time. Despite this variation, the way in which ritual efficacy is evaluated is constrained by intuitive casual principles. For example, first consider this ritual, taken from the Papyrus Ebers that was used to treat blindness, featured by Legare and Souza (2012): "Crush, powder, and make into one, the two eyes of a pig (remove the water therefrom), true collyrium (i.e., mineral eye salve), red-lead (i.e., red oxide), and wild honey [in a clay bowl]. Inject [mixture] into the ear of the patient. When thou hast seen properly to this mixing repeat this formula: 'I have brought this thing and put it in its place. The crocodile [god Sobek] is weak and powerless.' Repeat twice. Thereby he will at once recover" (*The Papyrus Ebers* 1931, 104). Now consider a ritual used to find a romantic partner in Brazil: "Buy a new sharp knife and stick it four times into a banana tree on June 12th at midnight.[1] Catch the liquid that will drip from the plant's wound on a crisp, white paper that has been folded in two. The dripping liquid captured on the paper at night will form the first letter of the name of your future partner" (Scharf 2010).

On the surface, there are many differences between the ritual described in the Papyrus Ebers and the ritual found in Brazil. For example, they involve the use of different substances (i.e., red-lead vs. sap from a banana tree), use different procedural steps (i.e., mixing vs. paper folding), use different artifacts (i.e., clay bowl vs. a knife), and are meant to address different problems (i.e., blindness vs. attracting a partner). At a deeper, conceptual level, however, there are many similarities. For one, they involve repetition of the procedural steps involved in the ritual (i.e., repeat twice vs. twice a day for two weeks), a large number of procedural steps (i.e., seven vs. six), specificity of time for when the actions should be performed (i.e., early rising vs. June 12th at midnight), procedural detail (i.e., mixing wild honey vs. buying a new sharp knife), and the invocation of supernatural agents (i.e., Sobek, an ancient Egyptian deity vs. Saint Anthony, a Catholic marriage saint).

Biases in causal reasoning are used to evaluate the efficacy of any type of action, yet we propose that their influence on action efficacy judgments is especially salient and influential when information about causal mechanisms is unavailable (Legare and Souza 2012, 2014). This rationale is based on a long-standing philosophical tradition that supports the claim that beliefs about causal connections arise from impressions of repeated instances of similar relations (Hume 1740). Below, we describe the causal biases associated with ritual action that may influence the perception of ritual efficacy:

(1) Repetition: The repetition of similar actions (e.g., pressing a button repeatedly to call an elevator) may be perceived to be causally efficacious because repetition makes behavioral information more psychologically available (Oppenheimer 2008), familiar (Scott and Dienes 2008), and attractive (Zajonc 1968).
(2) Number of steps: A larger number of procedural steps, over a smaller number of procedural steps, may increase the perception of causal efficacy by giving the impression that multiple actions may contribute to the production of a particular effect (Depue et al. 2006).
(3) Specificity: Given that human beings are good at perceiving or attributing intentions, seeing someone engaging in a detailed course of actions (e.g., catching the liquid that will drip from the plant's wound on a crisp, white paper that has been folded in two) may give the impression that particular details of the action sequence (i.e., time specificity, item specificity) have the potential to produce the desired outcome, even if the connection between the specific actions and the outcome is unknown or unavailable (Legare and Souza 2012, 2014).
(4) Supernatural agency: Intuitions about ritual efficacy may invoke the involvement of a supernatural agent at some level in the ritual sequence

(Barrett and Lawson 2001; Sørensen et al. 2006), in addition to intuitive causal principles (Sax et al. 2010). Involving supernatural agency in the ritual efficacy evaluation process may increase the likelihood that ritual actions are not perceived as bound by the same physical-causal expectations as non-ritualistic actions (Boyer 2001; Barrett and Malley 2007).

Given the role of supernatural agency in many rituals, there may be differences in the way believers (those who incorporate the supernatural into their worldviews) and nonbelievers, or those who do not adhere to or do not know about the specific ritual contexts and beliefs, reason about ritual efficacy. Even among believers, there are important cultural differences in terms of how people reason about supernatural agents and religious beliefs and practices. For instance, religious practices are emphasized more strongly than religious beliefs in Jewish relative to Christian traditions (Cohen et al. 2003). Other research has documented that even the scale of the society (small vs. large) may have important implications for how ritual is conceptualized: In societies with the belief in an all-knowing supernatural being, the primary concern is on acting morally as part of consensus with shared cultural beliefs (as often found in large-scale societies); in cultures in which supernatural beings are not concerned with, or privy to, social or moral knowledge, the emphasis is on performing costly rituals as a means of demonstrating socially shared behaviors (as often found in small-scale societies) (Purzycki and Sosis 2011). This suggests that the relative importance of religious behaviors (i.e., actions) versus mental states (i.e., beliefs) is also likely to play a role in reasoning about ritual efficacy.

Experimental research on evaluating ritual efficacy

Recent experimental research by Legare and Souza (2012) examined the "hidden logic" of ritual (Sax 2010). Rather than investigating the efficacy of ritual by examining outcomes or experience (Csordas 2002), they examined the kinds of information that influence perceptions of the efficacy of ritual action using experimental methodology to be able to make generalizations as well as draw causal inferences. Using ecologically valid content, they collected data in Brazil, a cultural context in which rituals or "recipes"—called *simpatias*—are available, endorsed, and used for everyday problem-solving purposes.

Simpatias are ritualistic remedial procedures and are not confined to any particular Brazilian religious group (even though some of them do include religious information). They are used to solve a variety of everyday problems (e.g., sinusitis, asthma, depression, anxiety, lack of luck, and infidelity). *Simpatias* are available to the general population, are relatively low-cost, and require no specialized expertise to be performed. Legare and Souza (2012) designed novel *simpatias*, modeled after content and information in real and widely available *simpatias*. The design of novel *simpatias* with ecologically valid content allowed for the experimental manipulation of the kind of information predicted to influence the evaluation of ritual efficacy (i.e., frequency of repetition, number of procedural steps, specificity of procedural detail, and presence of supernatural agents).

Four studies were conducted in this line of research. Study 1 was designed to develop and assess the ecological validity of the experimental stimuli. Study 2 examined potential kinds of intuitive information or *criteria* that may influence how ritual efficacy is evaluated. Study 3 provided a more systematic investigation of the intuitive criteria that were found to influence the evaluation of ritual efficacy in Study 2 and explored the impact of these criteria on reasoning about ritual efficacy among believers. Study 4 examined the extent to which the findings from Studies 1–3 represent universal features of human cognition, using identical stimuli in a cultural context unfamiliar with these ritualistic practices (US sample).

In Study 1, participants were presented with the experimental *simpatias* in written format and asked to evaluate whether or not they thought the *simpatia* would bring about the desired outcome. The results of this study demonstrated that nine criteria were potentially relevant to efficacy judgments: (1) specificity of time, (2) specificity of place, (3) specificity of material, (4) repetition of procedures, (5) number of procedural steps, (6) number of items used, (7) edibility (presence or absence of edible items), (8) digestibility (presence or absence of any sort of ingestion), and (9) religious icon (presence or absence of a religious icon).

Study 2 was designed to explore the extent to which these nine criteria affect the evaluation of ritual efficacy. Similar to Study 1, participants were presented with 18 experimental *simpatias* in written form that varied along the dimensions of frequency of performance, number of steps, and specificity of procedures. The results demonstrated that the rituals specifying greater frequency of performing the ritual act and the rituals specifying a greater number of specific steps or specifying a particular time the acts should take place were perceived as reliably more effective than rituals lower on these dimensions or lacking in this information. Study 3 used a sample of religious believers and replicated the findings of the previous study. Additionally, Study 3 also found evidence that rituals that included a religious icon were rated as significantly more effective than the ones without a religious icon. Study 4 replicated the same findings with a sample of US undergraduates that were unfamiliar with this type of ritual. The results of these studies support the hypothesis that the structure of ritual can be interpreted in light of intuitive causal beliefs about action efficacy.

To our knowledge, these were the first studies to investigate how ritual efficacy is evaluated from a psychological perspective. The results provided support for the proposal that information reflecting intuitive causal principles (i.e., repetition of procedures, number of procedural steps (Studies 2–4), and procedural specificity (specificity of time, Studies 2 and 3) increased evaluations of ritual efficacy. Repetition and a greater number of procedural steps might be activating biases in intuitive causal reasoning that evolved to reason about cause and effect relationships.

In addition to the effects of repetition, greater number of procedural steps, and procedural specificity, the results of Studies 3 and 4 demonstrated that *simpatias* that included a religious icon were perceived as more efficacious than *simpatias* without a religious icon. One possible explanation for this difference between Studies 2 and 3 may be that the participants in Study 3 were believers and that the participants in Study 2 included both believers and nonbelievers. The participants in Study 4 (university undergraduates) were unfamiliar with *simpatias*, yet appealing to religious idols (saints) for restorative or protective purposes is a common practice in Catholicism and thus may have been familiar to US participants. Thus, data from Studies 3 and 4 support the proposal that association with a supernatural agent (Barrett and Lawson 2001; Sørensen et al. 2006) impacts perceptions of ritual efficacy, especially for believers.

Familiarity with the content of the *simpatias* alone is not likely to account for these effects; the data do not support the possibility that any kind of familiar information (or information typically found in *simpatias*) increases perceptions of ritual efficacy. If familiarity alone were responsible for these effects, any kind of information frequently available in *simpatias* would influence the evaluation of ritual efficacy. Importantly, however, information about all 9 of the criteria tested in Study 1 are available in commonly used *simpatias*, and yet, the results of Study 2 demonstrated that information about where the *simpatia* should be performed, the number of items involved, where these items should come from or whether these items were edible or ingestible had no effect on ritual efficacy ratings. The core effects were also replicated with US undergraduates, a population that is entirely unfamiliar with the culturally specific content of *simpatias* (Study 4).

Whereas the number of experimental studies examining ritual has increased dramatically in recent years (Whitehouse 2001; Boyer and Liénard 2006), prior research on ritual cognition has not been based on real rituals that exist within particular cultural contexts and has focused almost exclusively on artificial or novel rituals (Barrett and Lawson 2001; Sørensen et al. 2006). The kind of ritual examined in these studies (*simpatias*) does not require specialized expertise, which allowed for the study of the evaluation of ritual efficacy experimentally using ecologically valid content. The use of culturally meaningful content by Legare and Souza (2012) to create experimental stimuli provides a novel methodological tool for investigation of ritual cognition.

Ritual and perceived control

Whereas the above studies demonstrated how ritual efficacy might be evaluated, until recently, little has been known about how manipulating perceived lack of control could influence perceptions of ritual practices. Examining the relationship between perceived control and ritual is of pervasive interdisciplinary interest with roots in both anthropology (Malinowski 1948; Boyer and Liénard 2006) and experimental psychology (Rudski and Edwards 2007; Kay et al. 2009). Malinowski (1948) proposed that in times of uncertainty, stress, and danger, people turn to magical rituals as a means of coping with stress and enacting some measure of control. When Malinowski visited the Trobriand Islands of New Guinea, for example, he noted that at times the Trobrianders would base their behavior on practices with clear causal rationales when fishing in a reliable and safe setting such as the lagoon; they described their successes and failures in terms of skill, whereas extensive ritual preceded the uncertain and dangerous conditions of deep-sea fishing. This behavior is not confined to the Trobriand Islands, up to 70 percent of college students in the United States employ ritualistic strategies to assist with performance on exams (Gallagher and Lewis 2001) or athletic competitions (Van Raalte et al. 1991; Cibrowski 1997; Vyse 1997; Bleak and Frederick 1998; Todd and Brown 2003).

There is also evidence that cultural ritual practices reduce stress during uncertain and dangerous times. For example, psalm recitation was successful in helping Israeli women cope with the stress of war (Sosis 2007). This suggests that "psalm recitation is likely to have emerged as an expected cultural norm during times of crises within Israeli communities because of its ability to buffer against the stress of uncontrollable conditions" (Sosis and Handwerker, 2011, 50). Psalm recitation may serve as a form of palliative coping with stressful, uncertain, and uncontrollable conditions. This work, as an updated version of Malinowski's theory of magical ritual, can be readily integrated with models of the mechanisms of ritualized behavior, such as that of Boyer and Liénard (2006).

Boyer and Liénard (2006) provide a model of ritualized behavior, accounting for behaviors seen in the Obsessive Compulsive Disorder spectrum, normal individual ritualized behaviors, and individual ritualized behaviors at different points in the lifespan (i.e., childhood, pregnancy, the early years of parenthood). The "Hazard Precaution System" is a psychological system geared toward dealing with inferred threats, as opposed to manifest threats encountered in the environment. Inferred threats, such as social exclusion, clues to the presence of predators or threatening conspecifics, and contamination are thought to activate mental "security systems" (Szechtman and Woody 2004; Boyer and Liénard 2006) that result in security-related behavior and coping strategies. Many collective rituals involve attention to potential danger and prescriptive, rigid behavioral patterns for averting the perceived danger and in this way resemble ritualized behaviors in OCD and childhood. As the authors note, "in collective rituals, people's insistence on the potential danger of not following the rules—expressed as moral reprimand (moral threat), as possible exposure to gossip or ridicule (threat of social exclusion), or as worry about misfortune—is very likely to activate the hazard-precaution system" (Liénard and Boyer 2006, 823).

In this model, anxiety generated by inferred threats focuses attention on low-level units of behavior instead of goal-related aspects of an action sequence, resulting in goal demotion. They argue that cultural rituals mimic the psychological system's normal inputs, making them attention grabbing and compelling. Ritual provides relief from anxiety through the "swamping" of working memory provided by reproducing actions that require a high level of cognitive control. Because anxiety is reduced through engagement in cognitively demanding actions, ritualized action cannot become automatic and routinized (Boyer and Liénard 2006). Engaging in security-related behavior such as ritual could also contribute to an increased sense of control (Hinds et al. 2010). These possibilities are not mutually exclusive and, to our knowledge, have not yet been examined experimentally. Future research should aim to determine the relationship between anxiety and sense of control in association with ritualized behaviors.

Most rituals have little or no actual bearing on the success of instrumental outcomes (Lobmeyer and Wasserman 1986), thus rituals are produced and maintained by an *illusion of control*, a phrase coined by Langer (1975). Research has found that an illusion of control is inferred when participants believe or respond as if contingencies between their behavior and the outcome exist, even if the outcomes are random (e.g., Alloy et al. 1981; Matute 1994). All dependent measures of the illusion of control reflect a belief that one's actions can influence an outcome that is, in fact, outside of their control.

Feelings of control promote increased self-esteem, optimism, and greater sense of agency (Scheier et al. 1994). Despite the benefits associated with feelings of control (Kofta et al. 1998), people frequently experience situations in which they lack the capacity to exert the control they desire. Believing that one cannot predict or influence future events (feeling a lack of control) contributes to depressive and pessimistic behavior when facing challenging situations (Fast et al. 2009). Attributional biases are activated and strategies are implemented to restore feelings of control when people are unable to influence and predict their environment (Underwood 1996; Weary and Jacobson 1997; Weary et al. 2001; Vaughn and Weary 2003). For example, when primed with feelings of lack of control, individuals detect correlations among random sets of stimuli that are presumably unrelated (Whitson and Galinsky 2008; Kay et al. 2009).

A common procedure used in experimental research on the illusion of control involves presenting participants with various buttons or levers and exposing them to noncontingent outcomes associated with pressing the buttons (Rudski 2001; Rudski and Edwards 2007). These studies have focused on participant variables (i.e., who is likely to engage in ritual or superstitious behavior), which are important in predicting and understanding behavior, however situational and contextual variables also play a crucial role.

Engaging in causally opaque practices such as rituals may seem to be a paradoxical means of increasing perceived control, yet rituals provide individuals with an opportunity to exert agency through action, thereby giving the illusion of increased control, this could also be related to anxiety reduction as described above (Kay et al. 2009). Priming randomness *may* increase the perception of a lack of control through the activation of attributional biases to detect a connection between action and outcome as a means of reestablishing feelings of control. The perception of a connection thus increases the evaluation of ritual efficacy.

Legare and Souza (2014) examined whether priming randomness (lack of control) affects the perception of the efficacy of rituals. Study 1 was conducted in Brazil, using the same rituals (i.e., *simpatias*) used by Legare and Souza (2012). Participants were first primed either with perceptions of randomness or with a neutral control. To prime lack of control through feelings of randomness, a previously validated task called The Scrambled Sentence Task—SST (Kay et al. 2010) was used. In this task, participants are asked to unscramble ten sets of five words each so that four of the words form a grammatical sentence and an extraneous word is dropped. For example, this five-word set "folder me behavior puzzle their" would become the four-word sentence "their behavior puzzles me," and the extraneous word "folder" is dropped. Five sets contain words related to randomness (e.g., puzzle, confusion) and the other five contain only neutral words. Following priming, participants were asked to evaluate the effectiveness of the eighteen *simpatias* used in the previous studies. Their data demonstrate that participants primed with randomness rated the rituals as more efficacious than participants primed with a neutral control. A possible explanation for this effect is that the manipulation of randomness activated a need to reestablish perceived control. Rituals may provide a mechanism for accomplishing this goal.

To explore the validity of the findings, Study 2 was conducted with a US sample using identical stimuli. As in Study 1, they directly compared the mean efficacy ratings of rituals in the randomness and control conditions. The US sample rated the rituals as less effective than did the Brazilian sample. Across samples, however, participants primed with randomness rated the rituals as more efficacious than participants in the control condition. These results demonstrate that even with unfamiliar content (as was the case with the US sample), priming randomness increased ritual efficacy evaluations, providing convergent support for the results of Study 1.

The results of both studies support the hypothesis that perceptions of ritual efficacy are influenced by the urge to regain a sense of control. This effect may occur because rituals present an opportunity for the perception of a potential connection between action and outcomes. Priming randomness may also change the perceived cost of performing the ritual versus the perceived benefit of the intended outcome. Future research should seek to elucidate the evolved function of this behavior by exploring the relationship between perceptions of control and actual control as they relate to ritualistic behavior. Research of this kind could provide a more nuanced picture of the proposed psychological mechanism and if the effect produces positive consequences in behavioral outcomes and is not limited to psychological states. For instance, there is research suggesting that ritualization of treatment administration appears to affect patient confidence, therefore increasing the healing power of the treatment (Miller et al. 2009; Kaptchuk et al. 2010).

There is widespread interdisciplinary interest in the relationship between perceived control and ritual (Humphrey and Laidlaw 1994; Whitehouse 2001; Csordas 2002; Bloch 2004; Boyer and Liénard 2006; Rudski and Edwards 2007; Sørensen 2007); however, the studies by Legare and Souza (2012, 2014) are the first to examine the relationship between a perceived lack of control and reasoning about the efficacy of ritualistic practices used by others. The research described in this section has provided experimental evidence that ritual may serve as a mechanism for reestablishing the perception of control and has provided insight into the cognitive foundations of the evaluation of ritual efficacy.

Conclusions

The objectives of this chapter were threefold. First, we provided a cognitive psychological account to explain the use of ritual for instrumental functions. Next, we provided a review of new psychological research on reasoning about the efficacy of ritual. Examining perceptions of ritual efficacy provides unique insight into the cognitive biases that constrain human behavior and account for cross-culturally recurrent practices. Finally, we provided evidence for the relationship between ritual and perceptions of control. The studies presented in this chapter examined the cognitive foundations and consequences of ritual experimentally. Examining the evaluation of ritual efficacy experimentally provides a means to explore the connection between some of the defining features of ritual and intuitive causal reasoning that may impact the evaluation of ritual efficacy. We also provided a potential explanation for why individuals are motivated to engage in ritual as means of exerting control over potentially dangerous and anxiety-provoking situations. Until recently, these connections were solely examined from a qualitative perspective. Future research in this area is needed to provide further evidence for the causal connections between perceptions of ritual efficacy, control, and the use of ritual for instrumental purposes.

The studies we described highlight the use of mixed-methodologies from a variety of disciplines to inform the examination of complex cognitive and social phenomena, such as ritual. The use of ecologically valid experimental paradigms maximizes the best possible trade-off between internal and external validities (Markman et al. 2009). The methods and materials used in these studies closely approximated the real-life cultural practices under investigation. This innovative interdisciplinary methodology is a core contribution of this research.

Ritual provides a uniquely informative context for studying causal cognition. Examining intuitive causal reasoning provides unique insight into the cognitive underpinnings of the evaluation of ritual efficacy. Studying ritual from a cognitive psychological perspective contributes to the body of research (Rudski 2001; Boyer and Liénard 2006; Kay et al. 2008, 2010) demonstrating that one of the functions rituals serve is to make the world seem more comprehensible, certain, and predictable.

Note

1 This is Valentine's Day in Brazil, a day before Saint Anthony's Day. Saint Anthony is considered the saint of marriages in Brazil.

Bibliography

Alloy, L. B., L. Y. Abramson, and D. Viscusi. 1981. Induced mood and the illusion of control. *Journal of Personality and Social Psychology* 41: 1129–1140.

Ashforth, A. 2001. An epidemic of witchcraft? The implications of AIDS for the post-apartheid state. In H. Moore and T. Sanders, eds, *Magical interpretations, material realities*, 184–225. London: Routledge.

Atkinson, Q. D. and H. Whitehouse. 2011. The cultural morphospace of ritual form: Examining modes of religiosity cross-culturally. *Evolution and Human Behavior* 32(1): 50–62.

Atran, S. and J. Henrich. 2010. The evolution of religion: How cognitive byproducts, adaptive learning heuristics, ritual displays, and group competition generate deep commitments to prosocial religion. *Biological Theory* 5(1): 18–30.

Barrett, J. and T. Lawson. 2001. Ritual intuitions: Cognitive contributions to judgments of ritual efficacy. *Journal of Cognition and Culture* 1: 183–201.

Barrett, J. and B. Malley. 2007. A cognitive typology of religious actions. *Journal of Cognition and Culture* 7: 201–211.

Bell, C. 1992. *Ritual theory, ritual practice*. New York: Oxford University Press.

Bleak, J. L. and C. M. Frederick. 1998. Superstitious behavior in sport: Levels of effectiveness and determinants of use in three collegiate sports. *Journal of Sport Behavior* 21: 1–15.

Bloch, M. 2004. Bringing ritual to mind: Psychological foundations of cultural forms. *Journal of the Royal Anthropological Institute* 10: 202–203.

Boyer, P. 2001. *Religion explained: The evolutionary origins of religious thought*. New York: Basic Books.

Boyer, P. and P. Liénard. 2006. Why ritualized behavior? Precaution systems and action-parsing in developmental, pathological and cultural rituals. *Behavioral and Brain Sciences* 29: 1–56.

Cibrowski, T. 1997. Superstition in the collegiate baseball player. *Sport Psychologist* 11: 305–317.

Cohen, A. B., J. L. Siegel, and P. Rozin. 2003. Faith versus practice: Different bases for religiosity judgments by Jews and Protestants. *European Journal of Social Psychology* 33(2): 287–295.

Cohen, E. 2007. *The mind possessed: The cognition of spirit possession in Afro-Brazilian religious tradition*. Oxford: Oxford University Press.

Cohen, E. and J. Barrett. 2008. When minds migrate; conceptualizing spirit possession. *Joural of Cognition and Culture* 8(1–2): 23–48.

Crowley, V. 1989. *Wicca: The old religion in the new age*. London: The Aquarian Press.

Csordas, T. J. 2002. *Body/meaning/healing*. New York: Palgrave Macmillan.

Depue, B. E., M. T. Banich, and T. Curran. 2006. Suppression of emotional and nonemotional content in memory effects of repetition on cognitive control. *Psychological Science* 17(5): 441–447.

Fast, N. J., D. Gruenfeld, N. Sivanathan, and A. D. Galinsky. 2009. Illusory control: A generative force behind power's far-reaching effects. *Psychological Science* 20: 502–508.

Gallagher, T. J. and J. M. Lewis. 2001. Rationalists, fatalists, and the modern superstitious: Test-taking in introductory sociology. *Sociological Inquiry* 71: 1–12.

Gellner, E. 1969. A pendulum swing theory of Islam. In R. Robertson, ed., *Sociology of religion: Selected readings*, 127–138. Harmondsworth: Penguin.

Goody, J. 1986. *The logic of writing and the organization of society*. Cambridge: Cambridge University Press.

Hinds, A., E. Woody, A. Drandic, L. Schmidt, M. Van Ameringen, M. Coroneos, and H. Szechtman. 2010. The psychology of potential threat: Properties of the security motivation system. *Biological Psychology* 85: 331–337.

Hume, D. 1740. *A treatise of human nature* (1967, ed.). Oxford: Oxford University Press.

Humphrey, C. and J. Laidlaw. 1994. *The archetypal actions of ritual*. Oxford: Oxford University Press.

Hutton, R. 1999. *The triumph of the moon: A history of modern pagan witchcraft*. Oxford: Oxford University Press.

Irons, W. 1996. Morality, religion, and human evolution. In W. M. Richardson and W. J. Wildman, eds, *Religion and science: History, methods, dialogue*, 375–399. New York: Routledge.

Kaptchuk, T. J., E. Friedlander, J. M. Kelley, M. N. Sanchez, E. Kokkotou, J. P. Singer, ... and A. J. Lembo. 2010. Placebos without deception: A randomized controlled trial in irritable bowel syndrome. *PloS One* 5(12): e15591.

Kay, A. C., D. Gaucher, J. L. Napier, M. J. Callan, and K. Laurin. 2008. God and the government: Testing a compensatory control mechanism for the support of external systems. *Journal of Personality and Social Psychology* 95: 18–35.

Kay, A., D. Moscovitch, and K. Laurin. 2010. Randomness, attributions of arousal, and belief in God. *Psychological Science* 12: 1–3.

Kay, A., J. A. Whitson, D. Gaucher, and A. D. Galinsky. 2009. Compensatory control: Achieving order through the mind, our intuitions, and the heavens. *Current Directions in Psychological Science* 18: 264–268.

Kofta, M., G. Weary, and G. Sidek. 1998. *Personal control in action*. New York: Plenum Press.

Langer, E. J. 1975. The illusion of control. *Journal of Personality and Social Psychology* 32: 311–328.

Lawson, E. T. and R. N. McCauley. 1993. Crisis of conscience, riddle of identity: Making space for a cognitive approach to religious phenomena. *Journal of the American Academy of Religion* 61(2): 201–223.

Legare, C. H. and S. A. Gelman. 2008. Bewitchment, biology, or both: The co-existence of natural and supernatural explanatory frameworks across development. *Cognitive Science* 32: 607–642.

Legare, C. H. and P. A. Herrmann. 2013. Cognitive consequences and constraints on reasoning about ritual. *Religion, Brain, and Behavior* 3: 63–65.

Legare, C. H. and A. L. Souza. 2012. Evaluating ritual efficacy: Evidence from the supernatural. *Cognition* 124: 1–15.

Legare, C. H. and A. L. Souza. 2014. Searching for control: Priming randomness increases evaluations of ritual efficacy. *Cognitive Science* 38: 152–161.

Liénard, P. and P. Boyer. 2006. Whence collective rituals? A cultural selection model of ritualized behavior. *American Anthropologist* 108(4): 814–827.

Lobmeyer, D. L. and E. A. Wasserman. 1986. Preliminaries to free throw shooting: Superstitious behavior? *Journal of Sport Behavior* 9: 70–78.

Malinowski, B. [1925] 1948. *Magic, science and religion*. Garden City, NY: Doubleday Anchor.

Markman, A. B., J. S. Beer, L. R. Grimm, J. R. Rein, and W. T. Maddox. 2009. The optimal level of fuzz: Case studies in a methodology for psychological research. *Journal of Experimental and Theoretical Artificial Intelligence* 21: 197–215.

Matute, H. 1994. Learned helplessness and superstitious behavior as opposite effects of uncontrollable reinforcement in humans. *Learning and Motivation* 25: 216–232.

McCauley, R. N. and E. T. Lawson. 2002. *Bringing ritual to mind: Psychological foundations of cultural forms*. Cambridge, MA: Cambridge University Press.

Miller, F. G., L. Colloca, and T. J. Kaptchuk. 2009. The placebo effect: Illness and interpersonal healing. *Perspectives in Biology and Medicine* 52: 518–539.

Norenzayan, A., I. Dar-Nimrod, I. G. Hansen, and T. Proulx. 2009. Mortality salience and religion: Divergent effects on the defense of cultural values for the religious and the non-religious. *European Journal of Social Psychology* 39: 101–113.

Norenzayan, A. and I. G. Hansen. 2006. Belief in supernatural agents in the face of death. *Personality and Social Psychology Bulletin* 32: 174–187.

Norton, M. I. and F. Gino. 2013. Rituals alleviate grieving for loved ones, lovers, and lotteries. *Journal of Experimental Psychology: General* 143(1): 266–272.

Oppenheimer, D. M. 2008. The secret life of fluency. *Trends in Cognitive Science* 12: 237–241.

Purzycki, B. G. and R. Sosis. 2011. Our Gods: Variation in supernatural minds. In F. Ulrich, ed., *Essential building blocks of human nature*, 77–93. New York: Springer-Verlag.

Rappaport, R. 1999. *Ritual and religion in the making of humanity*. Cambridge: Cambridge University Press.

Rossano, M. J. 2012. The essential role of ritual in the transmission and reinforcement of social norms. *Psychological Bulletin* 138(3): 529–549.

Rudski, J. 2001. Competition, superstition, and the illusion of control. *Current Psychology* 20(1): 68–84.

Rudski, J. M. and A. Edwards. 2007. Malinowski goes to college: Factors influencing students' use of ritual and superstition. *The Journal of General Psychology* 134(4): 389–403.

Ruffle, B. and R. Sosis. 2003. Religious ritual and cooperation: Testing for a relationship on Israeli religious and secular kibbutzim. *Current Anthropology* 44: 713–722.

Sax, W. S. 2004. Healing rituals: A critical performative approach. *Anthropology and Medicine* II: 293–306.

Sax, W. S. 2010. Ritual and the problem of efficacy. In W. S. Sax, J. Quack, and J. Weinhold, eds, *The problem of ritual efficacy*, 3–16. Oxford: Oxford University Press.

Sax, W. S., J. Quack, and J. Weinhold. 2010. *The problem of ritual efficacy*. Oxford: Oxford University Press.

Scharf, R. 2010, February 18. How to get a husband the Brazilian way. Retrieved from http://deepbrazil.com/2010/02/18/how-to-find-a-husband-the-brazilian-way/.

Scheier, M. F., C. S. Carver, and M. W. Bridges. 1994. Distinguishing optimism from neuroticism (and trait anxiety, self-mastery, and self-esteem): A reevaluation of the Life Orientation Test. *Journal of Personality and Social Psychology* 67: 1063–1078.

Schjoedt, U., J. Sorenson, K. L. Nielbo, D. Xygalatas, and J. Bulbulia. 2013. Cognitive resource depletion in religious interactions. *Religion, Brain, and Behavior* 3(1), 39–55.

Scott, R. and Z. Dienes. 2008. The conscious, the unconscious, and familiarity. *Journal of Experimental Psychology: Learning, Memory and Cognition* 35: 1264–1288.

Shore, B. 1996. *Culture in mind: Cognition, culture, and the problem of meaning*. New York: Oxford University Press.

Sørensen, J. 2007. *A cognitive theory of magic*. Lanham, MD: Rowman Altamira Press.

Sørensen, J., P. Liénard, and C. Feeny. 2006. Agent and instrument in evaluations of ritual efficacy. *Journal of Cognition and Culture* 6: 463–482.

Sosis, R. 2007. Psalms for safety: Magico-religious responses to threats of terror. *Current Anthropology* 48: 903–911.

Sosis, R. and C. Alcorta. 2003. Signaling, solidarity, and the sacred: The evolution of religious behavior. *Evolutionary Anthropology* 12: 264–274.

Sosis, R. and P. Handwerker. 2011. Psalms and coping with uncertainty: Religious Israeli women's responses to the 2006 Lebanon War. *American Anthropologist* 113(1): 40–55.

Sosis, R., H. C. Kress, and J. S. Boster. 2007. Scars for war: Evaluating alternative signaling explanations for cross-cultural variance in ritual costs. *Evolution and Human Behavior* 28(4): 234–247.

Souza, A. L. and C. H. Legare. 2011. The role of testimony in the evaluation of religious expertise. *Religion, Brain & Behavior* 2: 1–8.

Szechtman, H. and E. Woody. 2004. Obsessive-compulsive disorder as a disturbance of security motivation. *Psychological Review* 111(1): 111–127.

Tambiah, S. J. 1979. A performative approach to ritual. *Proceedings of the British Academy* 45: 113–169.

The papyrus ebers. 1931. (C. P. Bryan, Trans.). New York: D. Appleton and Company.

The papyrus ebers. 1937. (B. Ebbell, Trans.). Copenhagen: Levin & Munksgaard.

Todd, M. and C. Brown. 2003. Characteristics associated with superstitious behavior in track and field athletes: Are there NCAA divisional level differences? *Journal of Sport Behavior* 26: 168–187.

Turner, V. 1969. *The ritual process: Structure and anti-structure*. Piscatawa, NJ: Transaction Publishers.

Underwood, G. 1996. *Implicit cognition*. New York: Oxford University Press.

Van Raalte, J. L., B. W. Brewer, and C. J. Nemeroff. 1991. Chance orientation and superstitious behavior on the putting green. *Journal of Sport Behavior* 14: 41–50.

Vaughn, L. A. and G. Weary. 2003. Causal uncertainty and correction of judgments. *Journal of Experimental Social Psychology* 39: 516–524.

Vohs, K. D., Y. Wang, F. Gino, and M. I. Norton. 2013. Rituals enhance consumption. *Psychological Science* 24(9): 1714–1721.

Vyse, S. A. 1997. *Believing in magic*. New York: Oxford University Press.

Weary, G. and J. A. Edwards. 1994. Individual differences in causal uncertainty. *Journal of Personality and Social Psychology* 67: 308–318.

Weary, G. and J. A. Jacobson. 1997. Causal uncertainty beliefs and diagnostic information seeking. *Journal of Personality and Social Psychology* 73: 839–848.

Weary, G., J. A. Jacobson, J. A. Edwards, and S. J. Tobin. 2001. Chronic and temporarily activated causal uncertainty beliefs and stereotype usage. *Journal of Personality and Social Psychology* 81: 206–219.

Werbner, R. P. 1989. *Ritual passage, sacred journey: The process and organization of religious movement*. Manchester: Manchester University Press.

Whitehouse, H. 1995. *Inside the Cult: Religious innovation and transmission in Papua New Guinea*. Oxford: Oxford University Press.

Whitehouse, H. 2001. Transmissive frequency, ritual, and exegesis. *Journal of Cognition and Culture* 1: 167–182.

Whitehouse, H. 2004. *Modes of religiosity: A cognitive theory of religious transmission*. Oxford: Altamira Press.

Whitehouse, H. and J. A. Lanman. 2014. The ties that bind us. *Current Anthropology* 55(6): 674–695.

Whitson, J. and A. Galinsky. 2008. Lacking control increases illusory pattern perception. *Science* 322: 115–117.

Zajonc, R. 1968. Attitudinal effects of mere exposure. *Journal of Personality and Social Psychology Monographs* 9: 1–27.

6

Atheism, Inference, and Intuition

Kelly James Clark

Ripped from the headlines

Philosophers could save a lot of time by taking their cues from the news. Recent headlines declare that science has demonstrated the rationality of unbelief and the irrationality of belief: "Logic Squashes Religious Belief, A New Study Finds."[1] The venerable *Scientific American* proclaimed: "Losing Your Religion: Analytic Thinking Can Undermine Belief."[2] Philosophical writings, relying on work in the cognitive science of religion, are equally dismissive of rational religious belief. According to Daniel Dennett (2006), God is nothing but an evolutionarily induced figment of our imagination. Science writer, Richard Dawkins (2006), contends that God is a delusion. While God may run in our genes, unbelief, so it is claimed, is the product of rational reflection. A cognitive malfunction produces religious beliefs, while reason grounds unbelief. Unbelief, however, has gotten a free pass. Yet one wonders: if religious belief is induced by psychic dispositions or urges, why would unbelief be immune?

The cognitive faculties and processes involved in the production and sustenance of religious belief are typically the Agency Detection Device (ADD) and the Theory of Mind (ToM), which combine to generate, among many more mundane agential beliefs, god beliefs (Boyer 2001; Atran 2002; Barrett 2004; Murray 2009; Clark and Barrett 2010, 2011). Religious belief, according to the standard account, is culturally recurrent, natural, and non-inferential. What about unbelief? Recent studies argue that inferential thinking is associated with unbelief.[3] Would such studies, assuming they are validated, show that atheists are rational but theists are not? Would they demonstrate the rational superiority of atheism over theism? After presenting, analyzing, and then affirming the correlation between inferential thinking and unbelief, I will argue, through a discussion of philosophical intuitions, that they do not show the rational superiority of unbelief over religious belief.

Studies show?

If religious belief[4] is, by and large, culturally recurrent, natural, and intuitive (non-inferential), then we might expect unbelief to be, by and large, relatively rare, nonnatural and inferential.[5] If one typically finds oneself, through no inferential effort on one's own part, believing in God, one might need to reason one's way to unbelief. Consider an analogy with folk physics, which like religious belief is culturally recurrent, natural, and intuitive (McCauley 2011). Folk or naive physics is our unreflective, perceptual understanding of the physical world. Folk physics might include simple and true generalizations such as "Dropped rocks fall to the ground" and "Rocks thrown hard enough through windows will break them." And it might include commonsense statements that run contrary to contemporary physics, which postulates a host of unobservable entities such as atoms and photons. The movement from folk physics to contemporary physics required an enormous amount of inferential effort, effort sufficient to override at least some of our natural intuitions.[6] Contemporary physics, requiring abstract thinking and complicated mathematics, is deeply counterintuitive and contrary to what we observe. Belief in contemporary physical theories, then, requires inferential thinking. Likewise, the rejection of our very natural religious beliefs may involve inferential thinking.[7]

Just this sort of reasoning guided Will M. Gervais and Ara Norenzayan through a series of studies to determine the effect of inferential (what they called "analytic") thinking on religious belief and unbelief (Gervais and Norenzayan 2012). Because the aforementioned headlines above relied on their studies, I will consider them in some detail.[8] Gervais and Norenzayan offered a series of inferential prompts to determine their effect on religious belief and unbelief. They hypothesized that inferential thinking would override one's more natural and intuitive cognitive inclinations toward religious belief.

In the first study, using the Cognitive Reflection Test developed by Frederick (2005), they offered three problems. Their study will make more sense if you stop and think through your own response to the problems before proceeding to their analysis. The problems are as follows:

1. A bat and a ball cost $1.10 in total. The bat costs $1.00 more than the ball. How much does the ball cost? ____cents
2. If it takes 5 machines 5 min to make 5 widgets, how long would it take 100 machines to make 100 widgets?_____minutes
3. In a lake, there is a patch of lily pads. Every day, the patch doubles in size. If it takes 48 days for the patch to cover the entire lake, how long would it take for the patch to cover half of the lake?_____days

In each case, the quick and easy intuitive response is incorrect, and the more deliberate inferential response is correct.[9]

Participants were then measured with respect to religious belief and unbelief, responding to questions such as the following:

> In my life I feel the presence of the Divine
> It does not matter as much what I believe as long as I lead a moral life
> I believe in God
> I just do not understand religion
> God exists
> The devil exists
> Angels exist

Gervais and Norenzayan found that success on the Cognitive Reflection Test was negatively correlated with affirmations of religious belief; inferential thinking, they claimed, was negatively correlated with religious belief. So, in their terms, Gervais and Norenzayan concluded: "This result demonstrated that ... the tendency to analytically override intuitions in reasoning was associated with religious disbelief" (Gervais and Norenzayan 2012, 494).

A second set of studies involved unconscious primes, with a series of prompts designed to elicit inferential thinking. For sake of illustration, let us consider just one, the disfluency prime.[10] Disfluency Primes involved fuzzy fonts (prime sample font) rather than the large clear fonts of the control group (control sample font). Gervais and Norenzayan's claim is that having to figure out a fuzzy font engages inferential thinking in a way not required when reading large and clear fonts. The subjects again completed a measure of belief in God/religiosity. And again, Gervais and Norenzayan concluded that the set of studies reinforce the hypothesis that inferential processing decreases religious belief.

More recent studies affirm the hypothesis that if religious belief is more intuitive and non-inferential, then unbelief should be a product of inferential reasoning. Shenhav, Rand and Greene (2011) conducted a CRT study similar to that of Gervais and Norenzayan with over 800 participants (U.S. residents) with a median age of 33; intuitive responses were positively correlated with religious belief and inferential responses with unbelief. Their two other studies combine with this one to show a correlation between intuitive thinking and belief in God, on the one hand, and inferential (analytic) thinking and unbelief, on the other.

Pennycook et al. (2012) sampled over 200 people across the United States with a median age of roughly 35. They measured inferential thinking style (analytic cognitive style, ACS) in two ways, first with a variant of the Cognitive Reflector Test and second with Base-Rate Conflict (BRC) problems (problems that contain a conflict between a stereotype and probabilistic information). Since religious engagement is likely correlated with religious belief, they measured belief according to an individual's reported level of participation in, for example, church and prayer. They also measured religious beliefs through one's degree of belief in heaven, hell, miracles, afterlife, angels and demons, and an immaterial soul. Finally, they queried participants about what kind of God, if any, they believe in: answers ranged from theism, to deism, to atheism. While they produced many nuanced results, overall they affirmed the intuition that inferential thinkers are more likely than intuitive thinkers to be unbelievers. Their first study, for example, offered evidence of "an analytic [inferential] tendency to ignore or override initial intuitive responses" (339). They concluded that inferential processing decreases the likelihood of supernatural belief.

Rationality

Cognitive science of religion has apparently shown a correlation between intuitive thinking and religious belief, on the one hand, and between inferential thinking and unbelief, on the other. Does this, then, show that atheists are rational and that theists are not?[11] Before assessing rationality claims, we need some basic understanding of what rationality is. Any definition of rationality is fraught with peril. Yet, without some specification, ascriptions of rationality usually amount to little more than assertions of approbation (for one's own beliefs) or disapprobation (for those who disagree with one).[12] Rationality, as I define it, involves the proper use of the set of intellectual tools that human beings have for discovering the truth or of getting in touch with reality. Although rationality aims at the truth, it does not always reach its target (one can have rational but false beliefs); it seeks but cannot guarantee the finding of truth. Yet, the goal of being rational is to acquire true beliefs. Thus lightly armed, let us offer the following definition of rationality:

> A person p is epistemically rational with respect to belief b, if, in acquiring b, p is doing p's best to acquire true beliefs or get in touch with reality.

"Doing one's best" is admittedly vague; by it I preclude turning every normal human being into a fastidious evidence-seeking philosopher or a scientist. People have lives to live, different cognitive styles and capacities, different interests, and different demands. Moreover, this would confer prima facie rationality on p with respect to b. P might have, should have, or come to acquire beliefs that contradict or undermine b. In short, rationality is defeasible. If we should acquire evidence against a belief, we are rationally required to reassess the belief and even, in certain cases, the cognitive faculty that produces such beliefs. For example, suppose we have a disposition to believe that we are better than average. Evidence to the contrary could and should undermine the rationality of holding that belief, as well as the reliability of that cognitive faculty.

Doing one's best to acquire true beliefs involves the use of one's cognitive faculties, both intuitive and inferential. Some psychological studies of intuitive vs. inferential thinking valorize inferential reasoning (see Varga and Hamburger 2014). Philosophers, who trade in arguments, valorize inferential over immediately acquired beliefs. But recent work in epistemology suggests that human cognition is impossible without a deep and wide set of intuitive beliefs (Plantinga 1993; Greco 2000).

We could not acquire true beliefs about the world without a vast set of intuitive beliefs. For example, we intuitively and rationally believe that 2 + 2 = 4, that causing the suffering of innocents is wrong, that the future will be like the past, and that there is a world outside of our minds. Intuitive thinking produces pervasive, foundational, and even true beliefs; we simply could not believe much of anything without such foundational beliefs. Very likely, most of our deepest and most pervasive belief faculties produce beliefs that are intuitive—including memory, perceptual and moral faculties, on the one hand, and on the other hand, our dispositions to believe in the past or in the external world, or that the future will be like the past (Clark 1990; Spelke and Kinzler 2007; Clark and Barrett 2011). It is beyond the scope of this paper to do more than point to what seems obvious—we finite and fallible human believers can and must rely on a host of intuitively given beliefs. Such beliefs are prima facie rational (assuming one is doing the best one can to acquire the truth).[13]

Some of our rational beliefs are also inferential.[14] A jury might judge the guilt of a defendant after careful consideration of the evidence pro and con, a person might decide to buy a certain shampoo after reading several reviews, a scientist might slowly come to believe in descent with modification after a careful read of the *Origin of Species*, and Einstein might come to believe that $e = mc^2$ after some thought experiments and working out the mathematics.[15]

Rational beliefs, then, come in two forms—intuitive and inferential. As long as one is doing the best one can to acquire true beliefs, one's intuitively or inferentially acquired beliefs are prima facie rational.[16]

Even if we were to restrict rationality to inferential beliefs, we could not tell, based on these studies, whether or not any particular person was rational or irrational with respect to her belief or unbelief. Correlational studies, while showing fascinating generalities, are incapable of showing the grounds for or basis of any particular person's belief or unbelief. Even if men were, as a group, to score higher on math tests than women as a group, it is simply not true that every man is better at math than every woman. From the general truth about the groups, we cannot infer that any particular man is better at math than any particular woman. So, too, we cannot make any judgments about any individual based on the general tendencies of the group he or she falls into. We simply don't know if any particular man is good at math or not. And any particular theist may have come to her belief on the basis of rational reflection while any particular atheist may have simply but unconsciously "accepted" what her unbelieving parents told her. Believer and unbeliever alike could have or lack inferential grounds for their beliefs. So even assuming that intuitive thinking is irrational while inferential thinking is rational, we cannot tell, in any particular case, if a person is rational with respect to her belief or unbelief.[17]

Gervais and Norenzayan (2013) are aware that not all atheists are inferentialists (and not all theists are anti-inferentialists). They have identified, in addition to analytic (inferential) atheism, at least four additional types. Perhaps the most interesting is Mindblind atheism seen in individuals who lack the ability to mindread, that is, to process or cognize other minds (Norenzayan et al. 2012). Such individuals are usually high on the autism spectrum, lacking to various degrees what we have called Theory of Mind. The higher one is on the autism spectrum (that is, the less one is able to cognize persons), the less likely one is to believe in a personal God (see Clark and Visuri Forthcoming). If theistic beliefs involve belief in a divine person, those who lack the ability to cognize persons are likewise likely to lack belief in a divine person. Some atheists, which Gervais and Norenzayan calls "InCREDulous," simply lack adequate and relevant cultural inputs to form religious beliefs. Religions typically involve costly and observable Credibility Enhancing Displays (CREDs) such as fasting, tithing, chastity, and even martyrdom (Henrich 2009). Such "hard to fake" actions, which signal one's commitment to cooperate with in-group members, seem to contribute to the credibility and acceptability of a religion within a culture (Bulbulia 2004; Sosis 2005). In a culture lacking CREDS (e.g., in Northern European cultures), one might find religious belief simply incredible. Apatheism typically results from an indifference to God that arises from existential security; apatheists are content with their existential security and as a consequence do not care much about belief in God. Finally, apatheists care but they deeply want God not to exist. Jean-Paul Sartre, for example, did not want the kind of cosmic authority that an omniscient judge would have. Just as there are psychic, cultural, and non-truth conducive impulses involved in religious belief, so, too, there are psychic, cultural, and non-truth conducive impulses involved in unbelief. One cannot infer from the claim that atheists as a group have a more inferential thinking style that any particular atheist acquired her belief inferentially.

Both intuitive and inferential beliefs can be true; both can be rational. There is more than I can argue for here, but let me make the basic point. It is simply inconducive to our nature as human beings to restrict rationality to beliefs which can be inferred from some body of evidence. If we were to restrict ourselves to inferential beliefs, we would have nothing to believe (belief must start somewhere, not every belief can be inferred). If we have rational inferential beliefs, we must also have rational non-inferential beliefs.[18]

Inference and intuition

In the previous section, I claimed that inferential beliefs are based on beliefs that are ultimately intuitive. This is true, I think, in every domain of human inquiry. Our ordinary, commonsense beliefs rely on intuitions of space and time, the reliability of sense perception, belief in the past, and belief in an external world. Scientific beliefs assume without argument the uniformity of nature, the inductive principle, and truths of mathematics. In this section, I will focus on philosophical beliefs, in particular, belief in God. In so doing, I will appeal to the thought of William James, who lamented our failure to acknowledge the power of intuitions in human theorizing; as a result, he claimed that "there arises thus a certain insincerity in our philosophic discussions: the most powerful of all our premises is never mentioned" (James 1981, 9).

Reliance on intuition is often disguised by the remarkable complexity of philosophical arguments. Beneath the symbols, modalities, and nested propositions, one finds an intuition. In every philosophical argument, there is at least one fundamental premise that cannot be argued for. Dig deep enough, and one will find the unargued place where one starts. This unargued starting point is an intuition, an immediate, non-inferential judgment. Such intuitions may be elicited by stories, motivated by cases, critiqued by counterexamples, or appealed to in theories, but they are not and cannot be argued for. One "gets them" (or not).[19]

Although we must rely on our intuitions, we are not so metaphysically astute that we can clearly and certainly perceive those involved in, for example, an argument for (or against) the existence of God, for an absolute and universal moral standard, or for metaphysical idealism. Relevant intuitions in these fields might include claims that an infinite regress of causes is absurd, that moral statements require grounding, and that sensory appearances can be adequately accounted for without reference to a material world. Discussions in political theory, social policy, ethics, the meaning of life, the nature of human persons, determinism and free will likewise rely on crucial premises that are not universally discoverable by intuition. Widespread and interminable disagreement from epistemically equal peers is evidence against the indubitablity of philosophical intuition (McGinn 1993).

We all have substantive philosophical beliefs about reality, which belie commitments to fundamental, intuitive beliefs. Some substantive philosophical beliefs, for some or even most people, are immediately held, non-inferential; others are held on the basis of an argument. I take it that most ordinary folk hold few of their philosophical beliefs—in free will, say, or the objectivity of morality—on the basis of an argument and that, relatively speaking, more philosophers do. For many, including some philosophers, belief in God is immediate and non-inferential. And for others, perhaps for significantly more philosophers, belief in God is mediate and inferential. If the studies on atheism and inference are correct, relatively more unbelievers have come to their unbelief through inference. But everyone's belief, inferential or not, is grounded, ultimately, in intuition.

What are some of the intuitive beliefs involved in the epistemology of religious belief? With respect to basic believing, one might hold (or reject) the claims that belief in God is the sort of belief that must be based on evidence or that disagreement among those who are roughly our intellectual equals (in ability and information) undermines one's prima facie rationality. With respect to inferential belief, intuitive beliefs include the principle of sufficient reason, the objectivity of values (which factors into, among other things, the problem of evil and moral arguments for the existence of God) and the nature of inference (deductive, probabilistic, cumulative case?). But make no mistake: with either basic or inferential belief (or unbelief), one starts or ends with intuition.

Williams James holds that philosophical arguments are expressions of one's temperament. He claims that human inquiry, even in the sciences, inextricably involves one's "willing" or "passional" nature—one's temperament, needs, concerns, fears, hopes and passions. Every philosopher, James contends, "has taken his stand on a sort of dumb conviction that the truth must lie in one direction rather than another" (James 1956b, 93). We rely, in the end, on the way things seem to us, not on ineluctable intuition. We bring to philosophical argument our "dumb convictions"—pre-philosophical presuppositions about the way things seem. These seemings, according to James, are grounded in our sentimental, not our rational nature.

For James, these descriptive, psychological claims are philosophically necessary because of what has come to be called "the underdetermination of theory by data." Underdetermination holds that for any set of data, there are many hypotheses, which adequately explain the data but which are mutually incompatible with one another. Thus, the data alone cannot support one of these hypotheses over any other. As James puts it in his own earthy way:

> There is nothing improbable in the supposition that an analysis of the world may yield a number of formulæ, all consistent with the facts. In physical science different formulæ may explain the phenomena equally well ... Why may it not be so with the world? Why may there not be different points of view for surveying it, within each of which all data harmonize, and which the observer may therefore either choose between, or simply cumulate one upon another? A Beethoven string-quartet is truly, as some one has said, a scraping of horses's tails on cats' bowels, and may be exhaustively described in such terms; but the application of this description in no way precludes the simultaneous applicability of an entirely different description. (James 1956a, 76)

Our theories of the world—philosophical, religious, and even scientific—are underdetermined by the evidence. They are consistent with the data, but the data are not so compelling as to logically exclude competitors. When two such theories are under consideration, no appeal to the evidence could help us decide what to believe.

James suggests a method for deciding what to believe: "Well, of two conceptions equally fit to satisfy the logical demand, that one which awakens the active impulses or satisfies the æsthetic demands better than the other, will be accounted the more rational, and will deservedly prevail" (James 1956b, 76). "Suggests a method" would be too weak. James thinks that many philosophical options are forced (not to decide is to decide). Since the evidence underdetermines any of the options, one must rely on non-evidential ground to make the decision. Judgments about the nature of reality, then, require that we bring all that we are as human beings to bear on the matter at hand. He writes of philosophical reflection: "Pretend what we may, the whole man is at work when we form our philosophical opinions. Intellect, will, taste, and passion co-operate just as they do in practical affairs" (James 1956b, 92). Different persons with differing dumb convictions may find different "active impulses" awakened and engage or embrace entirely different intuitions (and thus entirely different conclusions from any argument based on those intuitions). We can and should expect that sincere and fully informed inquirers, insofar as they have different active impulses, may acquire radically different beliefs.

One might think that unacknowledged or deeply concealed intuitions were what James had in mind when talking about "the most powerful of all of our premises." But James rather looks beneath the premises of an argument to the person who holds them. Of philosophy, he writes to the effect that whoever touches an argument, "touches a man." We come to arguments contingently shaped, lacking a god's-eye view, and deeply limited in information. We have been shaped by evolution, genetics, culture, and friends to value this but not that, to see things this way but not that way, and to think in this style but not that one. We come to arguments as whole persons, not as disembodied souls or rational calculators, starting out at the very beginning with the "conviction that the truth must lie in one direction rather than another." Those sociohistorically and genetically shaped convictions are expressed in intuitions, which in turn shape one's beliefs.

One can make James's point by putting Plato and Aristotle side by side. Plato was deeply suspicious of sense perception, hoping to escape from this elusive and illusive shadowy world into the Real, ideal, and universal world of mathematics and the Good. Although deeply influenced by his teacher, Aristotle was constitutionally disposed to muck about, relish, and find reality in the very material world that Plato despised and to deny it to the world that Plato loved. Aristotle's philosophy affirms this world, particulars, and matter. While both argued for their particular worldviews, they relied fundamentally on different intuitions. Both could account equally well for all that humans experience. And yet, their conclusions were driven by their differing dumb convictions (intuitions) that the truth lies in this direction rather than that one. While their intuitions found expression in arguments, intuition not inference ultimately drove the development of their worldviews.

In philosophy, inferential and intuitive thinking are both grounded, ultimately, in intuition. Despite different styles, philosophical thinking is deeply and irremediably grounded in intuition.

Suppose James is right that our philosophical intuitions and thinking styles are rooted in nonalethic factors such as temperament. Add to this the growing body of experimental evidence which suggests that philosophical intuitions are likewise deeply affected by other nonalethic factors such as socioeconomic class and cultural background.[20] And suppose that James is right that philosophical arguments are based on such intuitions. Since philosophical arguments must rely on intuitions, neither resting on intuition nor relying on argument is better suited at gaining the truth.[21] With respect to philosophical matters (including belief or disbelief in God), then, intuition and inference are on epistemically equal ground. If rationality involves doing the best one can to get in touch with the truth, neither intuition nor inference has an epistemic advantage.

Conclusion

News headlines, some psychologists and many philosophers valorize inferential thinking over intuitive thinking. But human beings cannot avoid reliance on intuition. The situation is all the more pressing in matters philosophical. Scratch an inferentialist and you will find an intuitionist. That is, look carefully at a philosopher's proffered argument, and you will find an essential, intuitively accepted premise. Rationality must affirm intuitions. Even for the most ardent evidentialist, argumentative reasoning starts with intuitions. And while I think this is true for common sense and science, I think it more so in matters philosophical. That may be a bit too strong—intuition may not play more of a role in philosophy than, say, science, but in philosophy, intuitions are more deeply influenced by one's psychic makeup and culture than in science (hence widespread disagreement in philosophy).[22]

If one is doing the best one can with respect to gaining the truth, one's belief or disbelief in God is rational. Religious belief may be more nonreflective, but religious believers are not evidence insensitive. And atheists may be more inferential, but arguments assume intuitions. Finally, for all we know, atheism and theism alike may have deeply rooted psychic causes. The "for all we know" problem should give us pause: we simply do not know enough to make any normative, philosophical judgments about rational belief/unbelief based on our current state of knowledge in the cognitive science of religion.[23]

Notes

1 http://www.huffingtonpost.com/rob-brooks/analytic-thinking-erodes-religious-belief_b_1436940.html

2 http://www.scientificamerican.com/article.cfm?id=losing-your-religion-analytic-thinking-can-undermine-belief

3 These studies take analytic thinking as a synonym for inferential thinking (not as philosophers typically take it as relying on intuitive judgments). Since the intended audience of this essay is philosophers, I will not follow the psychologists and will instead use the term "inferential." I will remind the reader throughout of how I am using the terms.

4 Cognitive science of religion, while initially focused on *belief*, also concerns, for example, religious behaviors and communities. Moreover, religious belief itself is not well defined. In CSR, it might include belief in gods and ghosts, on the one hand, and benevolent and malevolent forces, on the other. For purposes of this paper, I take religious belief to refer, very roughly, to the high gods of, for example, Judaism, Islam, and Hinduism.

5 We are speaking in terms of general tendencies, not cognitive necessities. So, for example, while we (the entire group of human beings) may be generally inclined toward intuitive religious belief, not everyone will be a religious believer, and not every religious believer will have acquired his or her beliefs non-inferentially. The claim that we are typically natural and nonreflective theists is consistent with there being atheists and inferential theists.

6 Given our repeated relapses into folk physics, one might think that we can never fully overcome our natural dispositions.

7 Again, I am speaking in generalities. One might believe $e = mc^2$ because one was told it, not as a result of inferential thinking (though I doubt, under such circumstances, one would understand it well at all). Moreover, one might be an atheist because one's parents taught one at the earliest age that there was no God (and so required no inferential thinking on one's part).

8 Although Gervais and Norenzayan's studies prompted the headlines, they themselves resisted the sensational conclusions of the preceding section. They write: "Finally, we caution that the present studies are silent on long-standing debates about the intrinsic value or rationality of religious beliefs, or about the relative merits of analytic and intuitive thinking in promoting optimal decision making" (Gervais and Norenzayan 2012, 496).

9 The quick and easy intuitive yet wrong response to (1) is .10, while the correct analytic, deliberate answer is .05, to (2) is 100 while the analytic answer is 5 and to (3) is 24 while the analytic is 47.

10 The other studies involved implicit primes and art primes. Implicit Primes involved arranging words into sentences with the prime group given thinking terms (reason,

analyze, ponder, etc.), while the control group was given unrelated words (hammer, shoe, jump, etc.). Participants in the art control group stared at a "neutral" image such as *The Discobulos* whereas the remainder was primed by staring at *The Thinker* (an "artwork depicting a reflective thinking pose").

11 Some argue or imply, based on these studies, that atheists are more rational than theists or that atheists are rational and theists are not. Consider the ABC News headline on their story: "Religious faithfuls lack logic, study implies." http://abcnews.go.com/blogs/health/2012/04/26/logic-linked-to-religious-disbelief-study-implies/ Nicholas Epley claims that religious belief is "more of a feeling than a thought." http://news.sciencemag.org/2012/04/keep-faith-dont-get-analytical Faith is a matter of heart, unbelief of mind. In *Scientific American*, we read: "How Critical Thinkers Lose Their Faith in God" (theists, presumably, are uncritical thinkers) http://www.scientificamerican.com/article/how-critical-thinkers-lose-faith-god/ Rob Brooks, Scientia Professor of Evolutionary Ecology at UNSW Australia claims that it is becoming "clearer that religion is, in some senses, the opposite of rational thinking." http://theconversation.com/analytic-thinking-erodes-religious-belief-6709.

12 I take "rationality" to be the least stringent term of epistemic appraisal. Increasingly stronger terms of appraisal are "justified" and "warranted." On my account, most people are rational most of the time with respect to most of their beliefs, while considerably fewer are warranted (at least with respect to philosophical beliefs).

13 People have also intuitively believed that some people are naturally suited to be slaves, that the earth is flat and that women are not rational. Intuitive beliefs are not infallible.

14 Rationality as I understand it, as opposed to stronger terms of epistemic appraisal, is neutral with respect to whether or not the "evidence" is actual or putative. While actual evidence is clearly more truth-conducive, doing the best one can does not require one to ascertain in every or even most cases whether one's evidence is actual or putative. However, if one is made aware that one's evidence is either false or merely putative, one is in a different epistemic circumstance.

15 Or course, most of us do not know the evidence supporting $e = mc^2$, and so our rational belief that $e = mc^2$ is not based on evidence but on testimony.

16 I am assuming that it is people not beliefs that are rational, and so, the preferred locution would be that person *p* is rational with respect to a belief *b*.

17 It is, of course, a matter of debate whether or not belief in God is a rational non-inferential belief (Clark 1990; Plantinga 2000; Clark and Rabinowitz 2011). But this normative discussion is entirely independent of the empirical claim that inferential thinking is correlated with unbelief. Such empirical findings, while fascinating, are irrelevant to normative epistemic issues.

18 I am not insensitive to intuitive biases, which have been well documented (Kahneman 2011). But there are also inferential biases. For example, we tend to be sensitive to evidence or arguments which support our beliefs and to be insensitive

to evidence or arguments that are contrary to our beliefs. Not all inferential beliefs are true. People have inferred such untrue beliefs as the phlogiston theory, "women should aspire to be beautiful" (since they cannot be rational), and "Nixon will make a great president." Scientists seem to have inferred themselves into a contradiction between its two most widely accepted and successful theories—quantum mechanics and general relativity. They cannot both be true. Finally, philosophers, among the most ardent defenders of argument, continue to hold a wide diversity of incompatible beliefs. Some philosophers believe enthusiastically while others deny with equal vehemence the following and more (I take just a few claims in ethics; examples could be drawn from every area of philosophy): there are moral absolutes, there are moral facts, and there is human virtue.

19 Some philosophers contend that philosophical intuitions have evidential value, which others ardently reject (Cappelen 2012).

20 There is increasing empirical evidence that intuitions vary according to, for example, cultural background, socioeconomic status, and affective state (Weinberg et al. 2001; Nichols et al. 2003; Machery et al. 2004; Nichols and Knobe 2007; Swain et al. 2008).

21 One might think both are equally bad at gaining philosophical truth. Unlike many other intuitive beliefs, with philosophical intuitions we cannot check the facts to see if they are reliable. We have no belief-independent access to the philosophical world.

22 There may be an evolutionary story to tell here. It is hard to imagine a plausible evolutionary story in which developing reliable philosophical intuitions were reproductively successful.

23 I am grateful to Nicholas Wolterstorff, Alvin Plantinga, Aku Visala, and especially Ryan Nichols for helpful comments and criticisms.

Bibliography

Atran, S. 2002. *In Gods we trust*. New York: Oxford University Press.

Baron-Cohen, S., A. M. Leslie, and U. Frith. 1985. Does the autistic child have a "theory of mind"? *Cognition* 21: 37–46.

Barrett, J. 2004. *Why would anyone believe in God*? Walnut Creek, CA: AltaMira.

Boyer, P. 2001. *Religion explained: The evolutionary origins of religious thought*. New York: Basic Books.

Bulbulia, J. 2004. Religious costs as adaptations that signal altruistic intention. *Evolution and Cognition* 10: 19–38.

Caldwell-Harris, C., C. Murphy, T. Velazquez, and P. McNamara. 2012. Religious belief systems of persons with high functioning autism. Unpublished.

Cappelan, H. 2012. *Philosophy without intuitions*. Oxford: Oxford University Press.

Clark, K. J. 1990. *Return to reason*. Grand Rapids, MI: Eerdmans Publishing Company.
Clark, K. J. 2014. Atheism and analytic thinking. In *The science and religion dialogue: Past and future*. New York: Peter Lang Publisher.
Clark, K. J. and J. Barrett. 2010. Reformed epistemology and the cognitive science of religion. *Faith and Philosophy* 27(2): 174–189.
Clark, K. J. and J. Barrett. 2011. Reidian religious epistemology and the cognitive science of religion. *Journal of the American Academy of Religion* 79(3): 639–675.
Clark, K. J. and D. Rabinowitz. 2011. Knowledge and the cognitive science of religion. *European Journal of Philosophy of Religion* 3(1): 67–81.
Clark, K. J. and I. Visuri. Forthcoming. A. Coles and J. Collicutt. *The neurology of religion*. Cambridge: Cambridge University Press.
Dawkins, R. 2006. *The God delusion*. Boston, MA: Houghton Mifflin.
Dennett, D. 2006. *Breaking the spell: Religion as a natural phenomenon*. New York: Viking.
Dickson, L. 2010. Race and gender differences in college major choice. *The ANNALS of the American Academy of Political and Social Sciences, United Kingdom Sage Publishers* 627: 108–124.
Elqayam, S. and J. Evans. 2011. Subtracting "ought" from "is": Descriptivism versus normativism in the study of human thinking. *Behavioral and Brain Sciences* 34: 233–290.
Frederick, S. 2005. Cognitive reflection and decision making. *Journal of Economic Perspectives* 19: 25–42.
Gervais, W. M. and A. Norenzayan. 2012. Analytic thinking promotes religious disbelief. *Science* 336: 493–496.
Greco, J. 2000. *Putting skeptics in their place: The nature of skeptical arguments and their role in philosophical inquiry*. Cambridge: Cambridge University Press.
Henrich, J. 2009. The evolution of costly displays, cooperation, and religion: Credibility enhancing displays and their implications for cultural evolution. *Evolution and Human Behavior* 30: 244–260.
Henrich, J., S. J. Heine and A. Norenzayan. 2010. The weirdest people in the world? *Behavioral and Brain Sciences* 33(61): discussion 83.
James, W. 1956a. The sentiment of rationality. In *The will to believe and other essays in popular philosophy*. New York: Dover.
James, W. 1956b. *The will to believe and other essays in popular philosophy*. New York: Dover Publications.
James, W. 1981. *Pragmatism*. Indianapolis, IN: Hackett Publishing Company.
Kahneman, D. 2011. *Thinking fast and slow*. New York: Farrar, Straus and Giroux.
Lynn, R., J. Harvey, and H. Nyborg. 2009. Average intelligence predicts atheism rates across 137 nations. *Intelligence* 37(1): 11–15.
Machery, E., R. Mallon, S. Nichols, and S. Stich. 2004. Semantics, cross-cultural style. *Cognition* 92(3): B1–B12.
McCauley, R. N. 2011. *Why religion is natural and science is not*. Oxford: Oxford University Press.

McGinn, C. 1993. *The problems of philosophy: The limits of inquiry*. Hoboken, NJ: Wiley-Blackwell.

Murray, M. 2009. Scientific explanations of religion and the justification of religious belief. In: Jeffrey Schloss and Michael Murray, eds., *The believing primate: Scientific, philosophical, and theological reflections on the origin of religion*, 168–178. Oxford: Oxford University Press.

Nichols, S. and J. Knobe. 2007. Moral responsibility and determinism: The cognitive science of folk intuitions. Forthcoming in *Nous*.

Nichols, S., S. Stich, and J. Weinberg. 2003. Metaskepticism: Meditations in ethno-epistemology. In L. Stephen, ed., *The skeptics*, 227–247. Ashgate Publishing.

Norenzayan, A. and W. Gervais. 2013. The origins of religious disbelief. *Trends in Cognitive Sciences* 17(1): 20–25.

Norenzayan, A., W. Gervais, and K. Trzesniewski. 2012. Mentalizing deficits constrain belief in a personal God. *PLoS ONE* 7(5): e36880.

Pennycook, G., J. A. Cheyne, P. Seli, D. J. Koehler, and J. A. Fugelsang. 2012. Analytic cognitive style predicts religious and paranormal belief. *Cognition* 123(3): 335–346.

Plantinga, A. 1993. *Warrant and proper function*. Oxford: Oxford University Press.

Plantinga, A. 2000. *Warranted christian belief*. Oxford: Oxford University Press.

Shenhav, A., D. G. Rand, and J. D. Greene. 2011. Divine intuition: Cognitive style influences belief in god. *Journal of Experimental Psychology: General*. doi: 10.1037/a0025391.

Sosis, R. 2005. Does religion promote trust? The role of signaling, reputation, and punishment. *Interdisciplinary Journal of Research on Religion* 1: 1–30.

Spelke, E. and K. Kinzler. 2007. Core knowledge. *Developmental Science* 11: 89–96.

Stark, R. 2005. *What Americans really believe*. Waco, TX: Baylor University Press.

Swain, S., J. Alexander, and J. Weinberg. 2008. The instability of philosophical intuitions: Running hot and cold on truetemp. *Philosophy and Phenomenological Research* 76(1): 138–155.

Varga, A. L. and K. Hamburger. 2014. Beyond type 1 vs. type 2 processing: The tri-dimensional way. *Frontiers in Psychology* 5: 993. doi: 10.3389/fpsyg.2014.00993.

Weinberg, J., S. Nichols, and S. Stich. 2001. Normativity and epistemic intuitions. *Philosophical Topics* 29: 429–460.

7

How Do Philosophers Evaluate Natural Theological Arguments? An Experimental Philosophical Investigation

Helen De Cruz and Johan De Smedt

Introduction: Natural theology in philosophy of religion

Arguments for or against theism have an enduring appeal, and discussions about their merits continue to dominate debates in philosophy of religion. During the Middle Ages, God was a central topic of philosophical investigation, with authors like Anselm, Avicenna (Ibn Sīnā), and Aquinas studying God's existence and attributes through reason and arguments. Although religion no longer occupies the central place in philosophy it once held, it remains of lasting interest, with popular books on the topic of God's existence regularly making bestseller lists (e.g., Dawkins 2006; Keller 2008). Moreover, cosmological and design arguments appear in different cultures and times, for instance, in Greek and Roman antiquity, eighth- to tenth-century Hinduism, medieval Europe, and the Islamic world (see De Cruz 2014a for a review).

A large survey among professional philosophers (Bourget and Chalmers 2014)[1] reveals that theists are highly represented among philosophers of religion: 72.3 percent of philosophers of religion lean toward or accept theism, compared to 11.7 percent of philosophers outside of this field. This was the highest observed correlation between a philosophical specialization and a philosophical view ($r = .351$). Several nontheist authors (e.g., Levine 2000; Trakakis 2008; Draper and Nichols 2013) maintain that the overrepresentation of Christian theists in analytic philosophy of religion is unhealthy for the field, since they would be too much influenced by prior beliefs when evaluating religious arguments. However, a large percentage of theists does not, by itself, reveal that arguments in philosophy of religion are biased, nor does it specify what the influence of prior religious beliefs might be.

This study looks in more detail at the relationship between theism and the appraisal of natural theological arguments, that is, arguments that aim to establish theism or another metaphysical position through observation and reason. We aim to answer two interrelated questions: how philosophers appraise arguments for or against theism, and what factors influence their evaluation of such arguments. We will look in particular at the role of religious belief, philosophical specialization, and gender.

Methodology

The first author conducted an Internet survey. Participants rated eight arguments for theism and eight arguments against theism (Table 7.1) which were presented in a randomized order for each participant. Several of these arguments have different versions, for example, the Leibnizian and the kalām cosmological arguments. We presented the arguments just by their generic names, such as "ontological argument" and "cosmological argument." Participants were asked to rate how strong they thought each argument was on a scale from 1 to 5, with 1 = very weak and 5 = very strong. The prompt was "Please rate the following arguments according to how strong you find them." Arguments were presented in randomized order for each participant. Next to the ratings, participants could indicate if they were insufficiently familiar with the argument to rate it or

Table 7.1 Arguments presented to participants (order randomized for each participant) and their mean scores. The arguments are shown in descending order of perceived strength

Arguments for theism	Mean score
Cosmological argument	3.0
Argument from design	2.7
Argument from religious experience	2.7
Moral argument	2.5
Ontological argument	2.5
Argument from beauty	2.2
Pragmatic argument for theism	2.2
Argument from miracles	2.1

Arguments against theism	Mean score
Argument from evil	3.5
Argument from divine hiddenness	2.8
Argument from lack of evidence	2.8
Argument from parsimony	2.5
Argument from poor design	2.4
Argument from inconsistent revelations	2.4
Argument from incoherence	2.4
Pragmatic argument for atheism	2.1

decline to answer. Information about professional position, philosophical areas of specialization, age, gender, and religious belief was also collected. The options for religious belief were "theist," "atheist," or "agnostic/undecided" (henceforth referred to as "agnostic").

Results

Descriptive statistics of the participants

Participants ($N = 802$) were recruited through philosophy mailing lists (e.g., Philos-L) and three philosophy blogs (Prosblogion, NewApps, and Feminist Philosophers). These mailing lists and blogs are mainly read by professional philosophers and graduate students in philosophy who are the target population for this survey. The majority of respondents (82 percent) were professional philosophers, working as faculty members (32.9 percent), non-tenure track PhD holders, such as postdocs, visiting assistant professors, and adjuncts (15.8 percent), or graduate students (33.3 percent). The remaining respondents were undergraduates (8 percent) or not employed in academia (10 percent). Since these responses were not statistically different from those of the target population, we included them in the analysis. 24.2 percent of the respondents were female. Although the percentage of men in this sample is high, it is in line with the gender distribution among professional philosophers: the percentage of female philosophers in tenure-track or tenured positions in the United States is 19.8 percent, in the UK 22.3 percent, and in Canada 25.4 percent (see Buckwalter and Stich 2014 for an overview). The participants were on average 36.5 years old (SD = 11.8 years).

In this sample, 40.5 percent self-identified as theists, 40.4 percent as atheists, and 19.1 percent as agnostics. This distribution allows for robust comparisons between what people with different religious outlooks think about natural theological arguments. The percentage of philosophers of religion in this sample is also quite high. The most commonly reported areas of specialization[2] were philosophy of religion (33.8 percent), metaphysics (27.8 percent), ethics (26.8 percent), epistemology (25.8 percent), history of philosophy (22.2 percent), and philosophy of mind (19.2 percent).

Correlation between religious belief and specialization in philosophy of religion

We examined whether religious belief correlates positively with specializing in philosophy of religion. Among our respondents who listed philosophy of religion as an area of specialization, 73 percent were theists, 17 percent atheists, and 10 percent agnostics. Of those who did not list philosophy of religion as an area of specialization, 23.9 percent were theists, 52.4 percent atheists, and 23.7 percent agnostics. Philosophers of religion and nonphilosophers of religion significantly differed in their religious self-identification, $X^2(2, N = 802) = 180.359$, $p < .001$, Cramer's V = .474. There is a strong correlation between religious belief and philosophy of religion as an area of specialization ($r = .401$, $p < .001$).

While our findings replicate Bourget and Chalmers's (2014) percentage of theists in philosophy of religion (72.3 percent in their study, 73 percent here), we ended up with a higher percentage of theists among nonphilosophers of religion (11.7 percent in their study, 23.9 percent here). This may be the result of self-selection: given the subject matter, theists may have had more interest in completing this survey than nontheists. Alternatively, the higher percentage of theists in this study might reflect a more diverse demographic of faculty members and students. Sociological studies have found that atheism is more prevalent in elite universities than among regular faculty members (see, e.g., Gross and Simmons 2009). Bourget and Chalmers (2014) selected faculty members in leading PhD granting departments as their target participants, thereby excluding departments that are less prestigious, e.g., that only provide undergraduate or master degrees. Thus, they may have underestimated the percentage of theists in the population of professional philosophers due to their focus on elite institutions.

The percentage of theists among faculty members (tenured, tenure-track, and non-tenure track) was 36.6, whereas the percentage of theists in graduate and undergraduate students amounted to 45.9; the percentage of theists among nonacademic respondents was 37.5. Although this difference did not reach statistical significance, it is in line with data from Bourget and Chalmers (2014) which also indicate a higher percentage of theists among graduate students (20.9 percent), compared to faculty members (16.4 percent).

The perceived strength of religious arguments

Table 7.1 provides a summary of the arguments that were presented to the participants, with the mean score for each argument. Among the arguments for theism, those that were perceived as strongest were the cosmological argument (3.0), the argument from design (2.7), and the argument from religious experience (2.7). Among the arguments against theism, respondents judged the argument from evil to be the strongest (3.5), followed by the argument from divine hiddenness (2.8) and the argument from lack of evidence (2.8).

What explains the differences in ratings? One factor might be familiarity: an argument that receives more attention in the philosophical literature might be perceived as stronger. To examine this possibility, we took the total number of papers indexed for each argument in the PhilPapers database, one of the largest databases of philosophical works, containing not just journal papers but also monographs and articles in edited volumes. Since only two of the arguments against theism listed here were indexed, we concentrated on arguments for the existence of God in the survey, all of which were indexed except for the argument from beauty. The numbers of papers per argument are summarized in Table 7.2.

Table 7.2 Number of papers indexed in PhilPapers (accessed October 15, 2014) for each of the arguments for theism, in descending number of papers

Arguments for theism	Number of papers	Mean score
Religious experience	226	2.7
Cosmological argument	84	3.0
Pragmatic argument for theism	70	2.2
Ontological argument	69	2.5
Design argument	40	2.7
Moral argument	31	2.5
Argument from miracles	21	2.1

As can be gleaned from this dataset, there is no significant correlation between the perceived strength of the arguments and the number of papers that discussed them ($r = .39$, $p = .387$). Thus, familiarity does not predict the perceived strength of these arguments.

Religious belief predicts perceived strength of the arguments

We examined whether theists, atheists, and agnostics differ from each other in how they evaluate the natural theological arguments presented in this survey (see also De Cruz 2014b for a preliminary analysis of some of these results; and De Cruz and De Smedt 2015, 103–105, for a discussion of the results for the cosmological argument). We predict that philosophers evaluate arguments that support their prior beliefs as stronger than those that disconfirm their beliefs. Given that agnostics do not have a firm opinion on the matter, their beliefs would fall somewhere between those of theists and atheists.

A total score for arguments for theism was calculated by adding the individual scores for each of the eight arguments; the total score for arguments against theism is composed of the sum of all the individual scores for each of the eight arguments in that category.[3] Since each individual argument could be rated from 1 to 5, the maximum score an individual participant could reach for each type of argument was 40, that is, someone who rated all arguments for theism 5 would have a total score of 40 for the arguments for theism. The minimum score was 8, that is, someone who rated all arguments against theism 1 would have a total score of 8 for arguments against theism.

In line with our prediction, theists rated arguments that support theism significantly higher than atheists, whereas atheists rated arguments against theism significantly higher than theists. Agnostics occupy an intermediate position. The mean ratings for arguments for theism were 25.5 (SD = 5.7) for theists, 13.4 (SD = 5.8) for atheists, and 16.6 (SD = 5.2) for agnostics. The mean ratings for arguments against theism were 25.6 (SD = 6.9) for atheists, 17.7 (SD = 4.7) for theists, and 21.5 (SD = 5.8) for agnostics. A nonparametric test, the Kruskal–Wallis ANOVA, shows that these differences between theists, atheists, and agnostics are statistically significant, both for arguments for theism, $df(2) = 397.2$, $p < .001$ and for arguments against theism, $df(2) = 217$, $p < .001$. Effect sizes, calculated using post-hoc Mann–Whitney U tests, are moderate to large: for the difference between atheists and theists, $r = .78$ ($p < .001$); for the difference between atheists and agnostics, $r = .39$ ($p < .001$); and for agnostics and theists, $r = .58$ ($p < .001$). Figures 7.1 and 7.2 provide boxplots for the total scores for theism and against theism.

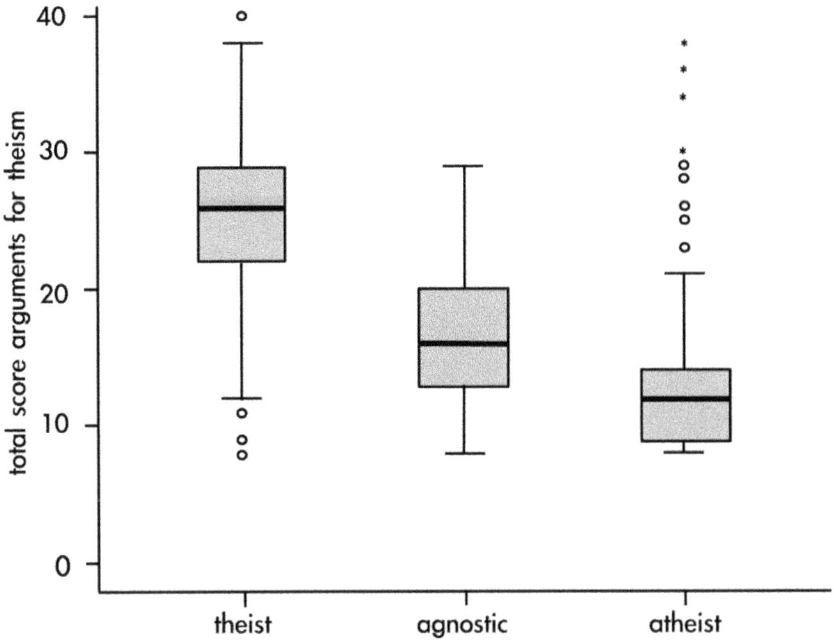

Figure 7.1 Boxplots with mean scores for all arguments for theism combined for theists, agnostics, and atheists (circles and stars denote outliers).

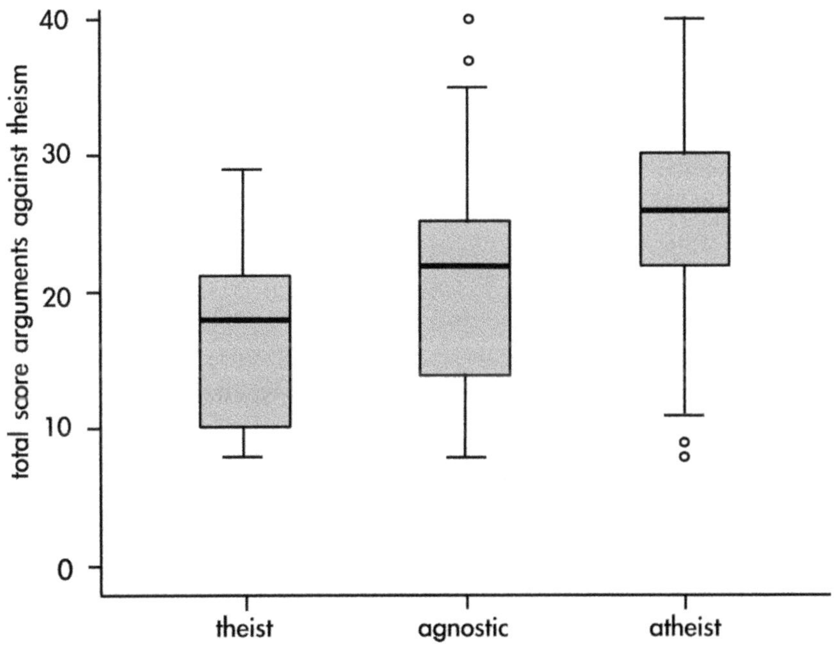

Figure 7.2 Boxplots with mean scores for all arguments against theism combined for theists, agnostics, and atheists (circles denote outliers).

Looking at the arguments separately, we find significant differences between the mean ratings of each argument by theists, atheists, and agnostics, with a greater difference of opinion about some arguments compared to others. We predicted that for the arguments where disagreement is largest, religious belief (atheism, theism, or agnosticism) correlates more strongly with how people evaluate these arguments. Since the opinions of theists and atheists are the most divergent, we took the mean difference between their ratings of individual arguments as a quantitative measure for the disagreement. Pearson's r indicates how the religious belief of participants correlates with their ratings.

For arguments in favor of theism, the cosmological argument elicited the largest disagreement between theists and atheists, with a mean difference of 2.08 (mean theists: 3.92, mean atheists: 1.85, $r = -.483$). The moral argument shows the second largest disagreement, with a mean difference of 2.01 (mean theists: 3.40, mean atheists: 1.38, $r = -.472$). Theists and atheists disagree least about the ontological argument and the pragmatic argument for theism: the mean difference for the former is 1.29 (mean theists: 3.01, mean atheists: 1.72, $r = -.346$), the mean difference for the latter is 0.95 (mean theists: 2.59, mean atheists: 1.64, $r = -.209$). All correlations are significant at the .001 level.

For the arguments against theism, theists and atheists disagree most strongly about the argument from lack of evidence, with a mean difference of 1.64 (mean theists: 2.16, mean atheists: 3.80, $r = .334$) and the argument from incoherence, with a mean difference of 1.58 (mean theists: 1.76, mean atheists: 3.34, $r = .389$). They disagreed least about the argument from evil and the argument from divine hiddenness. The former is the only argument where agnostics are not situated in between theists and atheists, as they rate it as less strong than either of the other groups (mean theists: 3.50, mean atheists: 3.76, mean agnostics: 3.2, mean difference between theists and atheists: .26). The correlation between religious belief and rating of the argument from evil is not significant ($r = -.013$). For the argument from divine hiddenness, the mean difference between theists and atheists is .43 (mean theists: 2.70, mean atheists: 3.13, $r = .055$). Here also, the correlation between religious belief and rating of this argument is not significant.

Looking at how strong the arguments are rated overall, there is an interesting asymmetry between the cosmological argument (rated as the top argument for theism) and the argument from evil (rated as the top argument against theism). Whereas theists and atheists disagree strongly about the strength of the cosmological argument, and their ratings of this argument can be predicted to a significant extent from their religious beliefs, the argument from evil does not elicit the same amount of disagreement. Indeed, there is no significant correlation between the religious beliefs (atheist, theist, and agnostic) of participants and their evaluation of this argument, and the difference between their mean ratings is small.

We thus found a confirmation for our prediction that religious belief significantly influences the evaluation of religious arguments: theists, atheists, and agnostics differ in how they evaluate arguments for and against the existence of God. The results are significant overall. For each of the arguments, except for the arguments from evil and from divine hiddenness, religious belief and rating of the argument correlate significantly.

Philosophical specialization influences the appraisal of some arguments

As we have seen, religious belief strongly predicts how participants evaluate natural theological arguments. Philosophers of religion gave significantly fewer "am unfamiliar with this argument" responses than nonphilosophers of religion for the following arguments: the ontological argument and the arguments from divine hiddenness, inconsistent revelations, poor design, and incoherence. The largest difference between specialists and nonspecialists was observed for the argument from divine hiddenness: 25.8 percent of nonphilosophers of religion were unfamiliar with this argument, compared to 4.4 percent of philosophers of religion ($X^2(1, N = 802) = 54.180, p < .001$, Cramer's V = .26).

Using a logit-probit model for the arguments for theism, and a cumulative logit model without proportional odds for the arguments against theism,[4] we investigated whether philosophical specialization has an effect that is independent from religious belief. In other words, taking into account that theists are disproportionately represented in philosophy of religion, does this specialization influence how philosophers evaluate the arguments? A cautionary note before proceeding with the statistical analysis of these results: as this is an exploratory study, where we are mainly interested in detecting potential patterns and less in avoiding false positives, we have not used Bonferroni or other methods of correction for multiple comparisons.[5] Given that we performed about fifty tests, the p-values would have had to be on the order of .001 or smaller to reach significance using Bonferroni correction.

Participants who have philosophy of religion as an area of specialization evaluated several arguments for theism more positively than those who do not. Philosophers of religion are 1.53 times more likely than nonphilosophers of religion to rate the cosmological argument favorably ($p = .01$). For the argument from design, philosophers of religion are 1.76 times more likely than nonphilosophers of religion to rate it favorably ($p < .001$). Philosophers of religion are 1.42 times more likely than other philosophers to rate the argument from miracles favorably ($p = .043$). They are also 1.39 times more likely than others to rate the argument from religious experience favorably ($p = .042$).

In this survey, the only argument against theism where philosophy of religion had a positive effect was the argument from divine hiddenness. Philosophers of religion were 1.54 times more likely than participants who do not have this area of specialization to rate this argument as "strong" versus "neutral," "weak," or "very weak" (p = .03). Philosophers of religion rated several arguments against theism as weaker. Nonphilosophers of religion are 1.82 times more likely than philosophers of religion to rate the argument from parsimony as "strong" versus "neutral," "weak," or "very weak" ($p = .02$). Nonphilosophers of religion are 2.26 times more likely than philosophers of religion to rate the pragmatic argument for atheism as "strong" versus "neutral," "weak," or "very weak" ($p = .02$). Subjects who are not philosophers of religion are 1.58 times more likely than those who are to rate the argument from incoherence as "neutral" versus "weak" or "very weak" ($p = .03$), and they are 1.76 times more likely than philosophers of religion to rate it as "strong" versus "neutral," "weak" or "very weak" ($p = .03$).

Gender effects

Only 9.6 percent of the respondents who were philosophers of religion were female, compared to 24.2 percent women in the entire sample. When focusing on faculty members (tenure-track, tenured, visiting assistant professors, and postdocs, $N = 391$), there were 12.3 percent women in philosophy of religion. The difference between the representation of women in philosophy of religion and in other philosophical specializations is statistically significant, X^2 (1, 391) = 11.507, p = .001, Cramer's V = .24.

In this survey, the percentage of theists among female respondents (22.2 percent) was significantly lower than the percentage of theists among male respondents (46.4 percent), $X^2(1, N = 802) = 35.78$, $p < .001$, Cramer's V = .24. To examine whether gender has an overall effect on the assessment of the arguments, we used a two-way ANOVA with religious belief and gender as independent variables. Controlling for religious belief, we found no difference, $F(11, 802) = 3.28$, $p = .071$ between how strongly male and female participants rated the arguments for theism combined and the arguments against theism combined.

Given the recent interest in the effects of gender on philosophical opinion (e.g., Buckwalter and Stich 2014; Seyedsayamdost 2015), we also examined whether gender is significantly correlated with assessing individual arguments. Male respondents found the cosmological argument and the argument from evil stronger. Female participants found the ontological argument, the pragmatic arguments for theism and atheism, and the arguments from inconsistent revelations and incoherence stronger.

In this survey, men were 1.52 times more likely than women to rate the cosmological argument more favorably ($p = .01$). They were 1.66 times more likely to rate the argument from evil as "strong" versus "neutral," "weak," or "very weak" ($p < .01$). Women were 1.43 times more likely than men to rate the ontological argument favorably ($p = .029$). They were 1.59 times more likely than men to rate the pragmatic argument for theism more favorably ($p < .01$). They were 1.53 times more likely than men to rate the argument from inconsistent revelations as "neutral" versus "weak" or "very weak" ($p = .02$); and 1.53 times more likely to rate it as "strong" versus "neutral," "weak," or "very weak" ($p = .03$). They were 1.81 times more likely than men to rate the argument from incoherence as "neutral" versus "weak" or "very weak" ($p < .01$), and 1.71 times more likely than men to rate the argument from incoherence as "very strong" versus "strong," "neutral," "weak," or "very weak" ($p = .04$). They were 1.75 times more likely than men to rate the pragmatic argument for atheism as "weak" versus "very weak" ($p = .01$), 2.27 times more likely to rate it as "neutral" versus "weak" or "very weak" ($p < .001$), 1.85 times more likely than men to rate it as "strong" versus "neutral," "weak," or "very weak" ($p = .01$), and 2.22 times more likely than male respondents to rate it as "very strong" versus "strong," "neutral," "weak," or "very weak" ($p = .03$). Remarkably, in this sample the arguments that men find stronger are also rated stronger overall, whereas several of the arguments that women find stronger are overall rated as weak—women also rate these arguments as relatively weak, but less so than men do.

For the argument from beauty and the argument from miracles, there is a significant interaction between gender and religious belief. Male theists were 26.83 times more likely than male atheists, and 7.72 times more likely than female atheists to rate the argument from beauty more favorably (both $p < .001$). They were 27.27 times more likely than male atheists, and 12.7 times more likely than female atheists to rate the argument from miracles more favorably (both $p < .001$). Male theists were 10.26 times more likely than male agnostics, and 3.56 times more likely than female agnostics to rate the argument from miracles more favorably (both $p < .001$). Female theists are 20.93 times more likely than male atheists and 9.75 times more likely than female atheists to rate the argument from miracles more favorably (both $p < .001$).

Discussion

This study has found several factors that significantly influence the appraisal of natural theological arguments for or against theism. The strongest predictor is religious belief, followed by philosophical specialization and gender. We will now discuss why these factors may play a role.

Confirmation bias

Several authors (e.g., Levine 2000; Trakakis 2008; Draper and Nichols 2013) have hypothesized that the religious beliefs of philosophers of religion (who are, to a large extent, Christian theists) unduly influence their appraisal of religious arguments. More specifically, theist philosophers of religion would evaluate the arguments in a selective way so that it supports their prior views.

People tend to evaluate evidence and arguments that are in line with their beliefs more favorably and to dismiss them when they do not support their beliefs. For instance, Taber and Lodge (2006) found that people consistently rate arguments in favor of their views on gun control and affirmative action more strongly than arguments that are incongruent with their views. When respondents could freely pick and choose information to look at, most of them actively sought out sympathetic, nonthreatening sources, for example, those in favor of gun control were less likely to read the sources against gun control that were presented to them and vice versa. Confirmation bias is the tendency to look for evidence that confirms, rather than disconfirms, one's earlier beliefs. It is accompanied by disconfirmation bias, the tendency to dismiss evidence or arguments for beliefs that are incompatible with one's own convictions (see Nickerson 1998; Mercier 2010 for review). These biases are not attenuated by education. For instance, molecular biologists—especially inexperienced ones—who find anomalous data tend to dismiss them as errors or experimental artifacts, rather than as potential disconfirming evidence (Fugelsang et al. 2004). The robustness of confirmation bias in highly educated people leads to the prediction that philosophers find arguments that support beliefs they already hold more persuasive than arguments that go against their prior beliefs.

Mere influence of prior beliefs on evaluation of arguments is indicative, but not conclusive evidence for confirmation bias. Suppose that philosophers all believe that 1 + 2 = 3 and are given an argument with a true mathematical conclusion that uses this as a central premise. The fact that philosophers come to believe the conclusion of this argument indicates that their prior beliefs influence their evaluation of the argument but is not evidence of confirmation bias. However, assume that some philosophers believe a controversial philosophical thesis (e.g., that free will is compatible with determinism), whereas others do no not. If the ratings by philosophers of arguments for or against compatibilism are significantly influenced by their prior beliefs, it would seem that confirmation bias plays a role. This would be especially the case if few philosophers changed their minds as a result of such arguments.

Our findings are consistent with a significant role of confirmation bias in philosophy of religion. Importantly, this bias is present in both theists and atheists. Atheists rate the arguments against theism more strongly than theists, and find the arguments for theism weaker than theists, and vice versa. Although being specialized in philosophy of religion makes a difference for the assessment of some arguments, religious belief remains the best predictor for how the arguments are rated.

When the first author reported preliminary findings of this survey on the philosophy blog Prosblogion, a number of philosophers expressed disapprobation at this significant role of confirmation bias, with one commenter writing "My reading of the situation is that philosophy of religion is unhealthy." Psychologists who study confirmation bias (e.g., Chinn and Brewer 1998) mostly regard it as having a negative impact on assessment and decision-making (hence the term "bias") as it leads people to ignore relevant evidence, to twist anomalous evidence to fit their expectations, and to fail to consider potentially viable alternatives. On the other hand, a disconfirmation strategy is often regarded as one that leads to sounder reasoning. This slipping of normative assumptions into psychological work is common (see Elqayam and Evans 2011 for review); it often has a significant impact on how empirical studies are designed. For instance, in confirmation bias studies, participants often do not receive alternative explanations, and working memory demands are heavy, which might make confirmation the only viable strategy. Using a carefully controlled framework in which participants were asked to use either a confirming or disconfirming strategy to evaluate evidence for a given theory (e.g., why whales get beached), Koslowski et al. (2013) found that disconfirmation does not automatically lead to a more careful assessment of potential sources of evidence than a confirmation strategy. In fact, participants who had to disconfirm a given hypothesis were more likely to think that evidence that was in fact neutral to the hypothesis disconfirmed it. As Koslowski et al. (2013) indicate, the term "confirmation bias" is actually a conflation of several forms of reasoning, running the gamut from a reasonable strategy (finding evidence to support a given hypothesis) to wishful thinking (deliberately ignoring potential disconfirming evidence).

Recently, proponents of the argumentative theory of reasoning (e.g., Mercier and Sperber 2011) have suggested that confirmation bias and other reasoning biases are not flaws, but that they are conducive to good reasoning. They are only flaws if we see reasoning as a solitary process of a detached, Cartesian mind. In order to reach a balanced conclusion, such a mind needs to be able to survey all the possibilities. However, when people are reasoning in dialogue with each other, such caution might not be necessary, and it might be more prudent to maintain one's belief in the face of disagreement if one thinks one is right: "In group settings, reasoning biases can become a positive force and contribute to a kind of division of cognitive labor" (Mercier and Sperber 2011, 73). Some studies (e.g., Trouche et al. 2014) indicate that people who are right are more likely to convince others in argumentative contexts than people who are more confident. In these studies, participants had to solve a puzzle with a nonobvious solution. Those who found the right answer tend to be better at convincing the others, because they have better-grounded arguments. Results like these indicate that even if individuals are biased, interaction through argumentation can lead to sounder reasoning. However, the participants in this study did not approach the puzzle with high or low prior probabilities attached to possible solutions, which is the case in the religious domain.

Even if the argumentative theory of reasoning is correct, it might be problematic that philosophers of religion evaluate the plausibility of theism using a confirmation strategy, which leads them to rate arguments in line with their prior beliefs more positively. Polarized debates in the public sphere, for example, about climate change, vaccines, and evolutionary theory indicate that the invisible hand of argumentative reasoning does not always work well, especially not in cases where one's own preconceptions play a significant role in the evaluation of arguments. This happens, for instance, if debaters are so entrenched in their views that they do not listen to their opponents, or even dismiss them out of hand as irrational (Morin 2014). Moreover, in cases where the invisible hand of argumentation works effectively—where biases at the individual level are corrected in argumentative contexts—it can only do so if there is a diversity of opinions. Thus, the large percentage of Christian theists in analytic philosophy of religion is somewhat worrisome. Among the atheist minority of philosophers of religion, the majority are naturalists. As Draper and Nichols (2013) and Schellenberg (in press) observe, it is striking that (Christian) theism and scientific naturalism are the only options that are systematically considered in philosophy of religion, to the extent that disconfirming evidence for one position is automatically seen as confirming evidence for the other: "most naturalists too assume that theistic God-centered religion must succeed if any does. Naturalism or theism. These seem to be the only options that many see. The harshest critics of religion, including philosophers such as Daniel Dennett, seem to think their job is done when they have, to their own satisfaction, criticized personalistic, agential conceptions of a divine reality" (Schellenberg in press). The role of confirmation bias in evaluating religious arguments can be mitigated by a broader pluralism in philosophy of religion.

Early-developed intuitions

We have argued elsewhere (e.g., De Cruz and De Smedt 2010, 2015; De Smedt and De Cruz 2011) that arguments in natural theology draw on intuitions and cognitive tendencies that arise developmentally early and robustly in human cognition. For example, the cosmological argument, which infers the existence of God from the existence of the universe, draws on causal intuitions we employ in everyday life. When an event occurs, we have the spontaneous intuition that something has caused it. Even infants and nonhuman animals have such causal intuitions. Experiments (e.g., Saxe et al. 2005; Newman et al. 2010) indicate that preverbal infants expect that something that happens, such as a stack of disordered blocks becoming ordered, has an external cause, and they prefer agents (e.g., a human hand) as causes over non-agents (e.g., a toy train).

The design argument relies on the intuition that ordered complexity is the result of purposive design. Features of the natural world, such as the eye or the cosmological constants and laws of nature, seem fine-tuned and designed for the functions they fulfill. A large, cross-cultural literature reveals that young children develop a strong preference for teleological explanations for the world they live in (see Kelemen 2004 for a review), and that these intuitions remain present throughout life.

The moral argument proposes that there are objective moral norms, and that the existence of such norms is more probable under theism than under naturalism. It capitalizes on our intuition that moral norms are objective, rather than purely contingent and subjective. From an early age onward, humans find moral conventions more objectively true than other norms, such as dress codes or taste preferences. For instance, young children think that hitting and pulling hair is not allowed in other countries, but they believe that dress codes may be more variable (Nichols and Folds-Bennett 2003).

The cosmological, design, and moral arguments depend on intuitions about causation, teleology, and ethics that are found in neurotypical members of our species; they are not specific to theism. This is in line with our empirical finding that these arguments are among the most highly rated arguments for theism. Remarkably, they also elicit a high degree of disagreement, with the cosmological and moral arguments eliciting the largest disagreement between theists and atheists.

If the premises on which these arguments depend rely on intuitions that are widespread, why do atheists and agnostics disagree about their conclusions? Religious arguments may be question-begging in a doxastic sense (Faust 2008). Although they are not strictly speaking circular (i.e., they do not contain their conclusions in their premises), reasoners accept their premises more readily if they also accept their conclusions. Thus, although it is intuitively plausible that moral norms are objective, nontheists tend to deny this premise of the moral argument and argue that moral norms are not really objective. Indeed, over the past decades, naturalists have offered detailed arguments for the claim that naturalism is incompatible with moral realism (e.g., Joyce 2006; Street 2006) and have also speculated why moral realism seems intuitively true (e.g., Ruse 2010). Similarly, although our everyday intuitions make the premise "whatever begins to exist has a cause for its existence" (the first premise of the *kalām* cosmological argument) plausible, nontheists routinely deny this premise and argue that the universe, unlike things in our everyday experience, caused itself to exist (Smith 1999).

Philosophical expertise

The existence of God is a classic case of peer disagreement where disagreeing parties are similarly educated and have access to the same body of evidence (perhaps with the exception of private evidence such as religious experience and *sensus divinitatis*), and it is therefore often discussed in the literature on peer disagreement (e.g., Frances 2014). Relevant evidence includes our intuitions about features of the world, such as (perceived) design and fine-tuning, the existence of the cosmos, moral norms, the pervasiveness of suffering among humans and other sentient beings, and the observation that God's existence is not an obvious fact (hiddenness). Several authors argue that keeping steadfast in one's own beliefs (e.g., van Inwagen 2010) is rational in the light of disagreements in natural theology. For instance, Rowe (1979) maintained that even in the face of particularly convincing cases of natural evil (such as a fawn dying slowly in a forest fire), the theist can still argue that this evil is not gratuitous, and therefore, that an atheistic conclusion does not follow. By contrast, conciliationists, such as Christensen (2007) and Elga (2007), have argued that one's confidence should at least be somewhat shaken when confronted with a disagreeing peer who has reached a different conclusion following the same evidence.

The majority of philosophers of religion are theists. Their opinions about God's existence differ significantly from philosophers who do not specialize in philosophy of religion, only a minority of which believes that God exists. Should this opinion of specialists carry some weight? What if one's interlocutor is an epistemic superior? Take Jake, an atheist epistemologist, who believes that the arguments from evil and divine hiddenness support his atheism. He does have some knowledge of philosophy of religion (mainly from the time of his undergraduate training), but he has never done a thorough study of the subject, let alone written any papers in philosophy of religion. Jake knows there are philosophers who have thought longer and harder about the question of God's existence than he did. On the basis of the same body of evidence, they have reached the opposite conclusion that God exists. Moreover, Jake knows that the majority of philosophers of religion have reached this same conclusion. Should he revise his beliefs or at least become less confident in his atheism?

In general, it seems sensible to defer to experts when one lacks the relevant expertise. If there is disagreement among experts, Goldman (2001, 97) recommends to look at the numbers and go with the majority, arguing that one can be "fully justified trusting E1 [an expert who defends the majority view] over E2 [an expert who is part of the minority] if almost all other experts on the subject agree with E1, or if even a preponderance of the other experts agree with E1." Given the high degree of consensus among philosophers of religion that God exists, the conclusion that one should at least accord weight to this consensus, or defer to it, seems to follow.

One can challenge this line of reasoning. Frances (2010) stipulates several cases in which one can blamelessly hold a philosophical position that is incompatible with what experts in the field think. For instance, the disagreement may be rooted in an undefended assumption or a factor where expertise does not extend. Philosophers of religion might be religious believers for other reasons than the arguments, such as their upbringing and feeling at home in a religious community which color their evaluation of evidence. Also, self-selection might play a role in who decides to specialize in philosophy of religion: philosophers who believe that God exists might be more motivated to study the field than those who do not. After all, if God exists, philosophy of religion is one of the most important areas of philosophy. But if he does not, it is merely on a par with the philosophy of other culturally widespread phenomena, like philosophy of music or philosophy of sport. Moreover, confirmation bias can explain to a large extent how philosophers evaluate natural theological arguments; this is the case for theists as well as atheists, for arguments for as well as against theism. A survey among philosophers (De Cruz in press) shows that rates of conversion in philosophers (since graduate school) are relatively small. Only 17.6 percent of philosophers reported a significant change in religious outlook, and for many of these, the change was not from theism to atheism or vice versa, but from one religious denomination to another. It seems that Jake can blamelessly believe that God does not exist while still holding that philosophers of religion, such as Eleonore Stump, Richard Swinburne, and Alvin Plantinga, are his epistemic superiors in particular subjects in philosophy of religion.

Jake might reach the opposite conclusion: the reason that the majority of philosophers of religion believe theism is true has little or nothing to do with their philosophical work. They already hold theism from the outset, and their theistic beliefs color their evaluation of the evidence. If anything, the fact that the majority of philosophers working in this field are theists points to a large role of irrelevant causal factors in reaching philosophical conclusions, which may further lower Jake's confidence in the soundness of arguments for theism. When authors like Draper and Nichols (2013) and Levine (2000) argue that philosophy of religion lacks vitality, they have this worry in mind.

Whether their misgivings are justified depends on whether permissivism is plausible. According to the uniqueness thesis (e.g., White 2005), there is only one rational attitude one can hold in the face of a total body of evidence, whereas permissivists (e.g., Kelly 2014; Schoenfield 2014) propose that there is some leeway, and that people can hold several positions in the light of the same evidence. There are different flavors of permissivism. Strong versions hold that one's beliefs are reasonable when they cohere with one's earlier beliefs (Schoenfield 2014), whereas moderate permissivists argue that only a certain range of beliefs is reasonable (Kelly 2014). From the perspective of uniqueness, only some philosophers are rational. For instance, according to Feldman (2007), suspending judgment (agnosticism) is the only rational option. For permissivists, philosophers can reasonably hold religious beliefs (including atheism) if they have good arguments for them; they can thus be significantly influenced by irrelevant causal factors and yet be rational. Take Jenny, who has been raised in an agnostic environment, by parents and teachers who never talked about God.[6] When she considers God's existence, her epistemic situation differs from that of other philosophers who were raised as theists or atheists, as she was exposed to neither firm religious beliefs nor disbeliefs. She lacks experiences and habits of mind that theists or atheists hold. The fact that her background differs from theirs has nothing to do with the truth or falsity of theism and is thus epistemically irrelevant; it is a matter of chance that she happened to grow up in an agnostic environment. If theist philosophers are influenced by irrelevant causal factors, so are atheists, and even agnostics like Jenny—no one comes from a neutral background that would yield an epistemically privileged perspective. Given the importance of irrelevant influences in shaping religious views, even if all philosophers of religion subscribed to theism, this would still not mean theism is true. The high percentage of theists in philosophy of religion is therefore not an indication of the truth of theism. (For analogous reasons, the low percentage of theists in the general philosophy population is not epistemically relevant either.)

While Jake may rationally hold his atheistic beliefs in the face of the large percentage of theists in philosophy of religion, he might still learn something, not about the overall conclusion one should draw from natural theological arguments, but about their individual strengths. As we saw, philosophy of religion has an independent effect on the perceived strength of some arguments. The cosmological argument and the arguments from design, religious experience, miracles, and divine hiddenness are rated as stronger by philosophers of religion than by other philosophers when controlling for religious belief. This may be a result of familiarity: nonphilosophers of religion are not familiar with the recent literature. While they know the argument from miracles vaguely through Hume's ([1748] 2007) dismissive discussion, they have little knowledge of recent scholarship, for example, probabilistic versions of the argument that take into account biblical research (McGrew and McGrew 2009). Thus, although our survey did not reveal a lower self-reported knowledge of the argument from miracles by nonphilosophers of religion, they are likely unfamiliar with the more recent versions. Likewise, the argument from divine hiddenness is treated in specialist literature (the majority of which appears in monographs, edited volumes, and philosophy of religion journals), with which nonphilosophers of religion are unfamiliar. This was revealed in the self-reported familiarity with the argument, with only 4.4 percent of philosophers of religion being unfamiliar with the argument, versus 25.8 percent of other philosophers. Since differences in the evaluation of the arguments remain when controlling for religious belief, it seems that familiarity with the recent specialist literature has some epistemic weight.

Gender

Like most other areas of philosophy, philosophy of religion exhibits a gender imbalance. In this sample, only 9.6 percent of philosophers of religion were female, whereas in the sample as a whole, 24.2 percent were women. Buckwalter and Stich (2014) propose that the underrepresentation of women in philosophy is partly caused by gender differences in philosophical intuitions. They provide several lines of purported evidence for gender differences in intuitions about philosophical scenarios, such as Gettier cases and Thomson's violinist. In this study, the cosmological argument and the argument from evil are rated as stronger by men, so it is unsurprising that they have an overall stronger rating, given the high percentage of male respondents in the sample. Some of the arguments that women find relatively stronger, for example, the pragmatic arguments for theism and atheism, and the arguments from inconsistent revelations and incoherence, are overall rated as weaker.

Buckwalter and Stich (2014) is in part a meta-analysis, which relies among others on responses collected from researchers and instructors who reported gender differences. That part of their study may therefore be subject to selection bias—researchers who obtained null results may have been less inclined to report their findings. As it stands, several papers that looked for gender effects in philosophical intuitions (e.g., Adleberg et al. 2015; Seyedsayamdost 2015) failed to replicate their results. In this study, we were mainly interested in detecting potential effects rather than minimizing type 1 errors, so we did not correct for multiple tests, which may have colored our findings on gender effects. They should be replicated to establish whether they live up to scrutiny.

Concluding remarks

This paper presents findings of a large quantitative survey on how philosophers evaluate arguments for or against the existence of God. To our knowledge, this is one of the first studies in experimental philosophy of religion. It is also the first to provide a quantitative measure of the role of religious belief in evaluating natural theological arguments and, more tentatively, of the role of philosophical specialization and gender. We replicated Bourget and Chalmers's

(2014) observation that a large majority of philosophers of religion are theists. We found a robust correlation between perceived strength of natural theological arguments and religious belief. Furthermore, we found a potential influence of philosophy of religion as an area of specialization and of gender on the evaluation of these arguments.

To further an understanding of the ways in which philosophical work and religious belief interact, qualitative data are also needed.[7] Such data can shed light on the role of religious upbringing and the relationship between privately held religious beliefs and publicly defended philosophical viewpoints. While we have focused on the appraisal of natural theological arguments, philosophy of religion is a much broader field, with areas such as concepts of God and the metaphysics of persons (e.g., body-soul dualism). Within cognitive science of religion, there is an increasing body of literature on these topics, but at present, there is little awareness of this among philosophers.

Acknowledgments

We would like to thank Ryan Nichols, Victoria Harrison, and Klaas Kraay for comments to this paper. This project was supported by a postdoctoral research grant from the British Academy (grant pf130006) and by a postdoctoral grant from Ghent University.

Notes

1 Bourget and Chalmers (2014) is the only other quantitative survey to date that probes the religious views of professional philosophers.
2 Among philosophers, it is common to have more than one area of specialization.
3 There was moderate to high internal consistency in the answers of respondents: Chronbach's alpha was .886 for arguments for theism and .749 for arguments against theism. Standard deviations of the responses to individual arguments were quite similar, ranging between 1.2 (argument from miracles) and 1.4 (cosmological argument).
4 We thank Robert O'Brien for calculating these statistics.
5 The Bonferroni correction is in many cases too conservative: it minimizes the risk of type 1 (false positive) errors but as a result may miss many significant results. Note that even using this conservative measure, the effect of religious belief on appraisal of the arguments remains present.

6 We do not mean to suggest that the agnostics in our sample come from a background like Jenny's. We provide this example as a supposedly ideal epistemic situation for questions about the existence of God.
7 One of us is currently working on a paper that addresses this question, using a qualitative survey.

Bibliography

Adleberg, T., M. Thompson, and E. Nahmias. 2015. Do men and women have different philosophical intuitions? Further data. *Philosophical Psychology* 28: 615–641.

Bourget, D. and D. J. Chalmers. 2014. What do philosophers believe? *Philosophical Studies* 170: 465–500.

Buckwalter, W. and S. Stich. 2014. Gender and philosophical intuition. In J. Knobe and S. Nichols, eds, *Experimental philosophy: Volume 2*. 307–316. Oxford: Oxford University Press.

Chinn, C. A. and W. F. Brewer. 1998. An empirical test of a taxonomy of responses to anomalous data in science. *Journal of Research in Science Teaching* 35: 623–654.

Christensen, D. 2007. Epistemology of disagreement: The good news. *Philosophical Review* 116: 187–217.

Dawkins, R. 2006. *The God delusion*. Boston, MA: Houghton Mifflin.

De Cruz, H. 2014a. The enduring appeal of natural theological arguments. *Philosophy Compass* 9: 145–153.

De Cruz, H. 2014b. Cognitive science of religion and the study of theological concepts. *Topoi. An International Review of Philosophy* 33: 487–497.

De Cruz, H. in press. Religious disagreement: An empirical study among academic philosophers. *Episteme*.

De Cruz, H. and J. De Smedt. 2010. Paley's iPod. The cognitive basis of the design argument within natural theology. *Zygon: Journal of Religion and Science* 45: 665–684.

De Cruz, H. and J. De Smedt. 2015. *A natural history of natural theology. The cognitive science of theology and philosophy of religion*. Cambridge, MA: MIT Press.

De Smedt, J. and H. De Cruz. 2011. The cognitive appeal of the cosmological argument. *Method and Theory in the Study of Religion* 23: 103–122.

Draper, P. and R. Nichols. 2013. Diagnosing bias in philosophy of religion. *The Monist* 96: 420–446.

Elga, A. 2007. Reflection and disagreement. *Noûs* 41: 478–502.

Elqayam, S. and J. S. B. Evans. 2011. Subtracting "ought" from "is": Descriptivism versus normativism in the study of human thinking. *Behavioral and Brain Sciences* 34: 233–248.

Faust, J. 2008. Can religious arguments persuade? *International Journal for the Philosophy of Religion* 63: 71–86.

Feldman, R. 2007. Reasonable religious disagreements. In L. Anthony, ed., *Philosophers without gods*, 194–214. Oxford: Oxford University Press.

Frances, B. 2010. The reflective epistemic renegade. *Philosophy and Phenomenological Research* 81: 419–463.

Frances, B. 2014. *Disagreement*. Cambridge: Polity Press.

Fugelsang, J. A., C. B. Stein, A. E. Green, and K. N. Dunbar. 2004. Theory and data interactions of the scientific mind: Evidence from the molecular and the cognitive laboratory. *Canadian Journal of Experimental Psychology* 58: 86–95.

Goldman, A. I. 2001. Experts: Which ones should you trust? *Philosophy and Phenomenological Research* 63: 85–110.

Gross, N. and S. Simmons. 2009. The religiosity of American college and university professors. *Sociology of Religion* 70: 101–129.

Hume, D. [1748] 2007. *An enquiry concerning human understanding* (P. Millican, ed.). Oxford: Oxford University Press.

Joyce, R. 2006. *The evolution of morality*. Cambridge, MA: MIT Press.

Kelemen, D. 2004. Are children "intuitive theists"? Reasoning about purpose and design in nature. *Psychological Science* 15: 295–301.

Keller, T. 2008. *The reason for God. Belief in an age of scepticism*. London: Hodder & Stoughton.

Kelly, T. 2014. Evidence can be permissive. In M. Steup, J. Turri, and E. Sosa, eds, *Contemporary debates in epistemology*, 298–312. Oxford: Oxford University Press.

Koslowski, B., J. Marasia, F. Vermeylen, and V. Hendrix. 2013. A disconfirming strategy is not necessarily better than a confirming strategy. *American Journal of Psychology* 126: 335–354.

Levine, M. P. 2000. Contemporary Christian analytic philosophy of religion: Biblical fundamentalism, terrible solutions to a horrible problem, and hearing God. *International Journal for Philosophy of Religion* 48: 89–119.

McGrew, T. and L. McGrew. 2009. The argument from miracles: A cumulative case for the resurrection of Jesus of Nazareth. In W. L. Craig and J. P. Moreland, eds, *The Blackwell companion to natural theology*, 593–662. Chichester: Wiley-Blackwell.

Mercier, H. 2010. The social origins of folk epistemology. *Review of Philosophy and Psychology* 1: 499–514.

Mercier, H. and D. Sperber. 2011. Why do humans reason? Arguments for an argumentative theory. *Behavioral and Brain Sciences* 34: 57–74.

Morin, O. 2014. The virtues of ingenuity: Reasoning and arguing without bias. *Topoi. An International Review of Philosophy* 33: 499–512.

Newman, G., F. Keil, V. Kuhlmeier, and K. Wynn. 2010. Early understandings of the link between agents and order. *Proceedings of the National Academy of Sciences USA* 107: 17140–17145.

Nichols, S. and T. Folds-Bennett. 2003. Are children moral objectivists? Children's judgments about moral and response-dependent properties. *Cognition* 90: B23–B32.

Nickerson, R. S. 1998. Confirmation bias: A ubiquitous phenomenon in many guises. *Review of General Psychology* 2: 175–220.

Rowe, W. L. 1979. The problem of evil and some varieties of atheism. *American Philosophical Quarterly* 16: 335–341.

Ruse, M. 2010. The biological sciences can act as a ground for ethics. In F. J. Ayala and R. Arp, eds, *Contemporary debates in philosophy of biology*, 297–315. Chichester: Wiley-Blackwell.

Saxe, R., J. Tenenbaum, and S. Carey. 2005. Secret agents: Inferences about hidden causes by 10- and 12-month-old infants. *Psychological Science* 16: 995–1001.

Schellenberg, J. L. 2016. Divine hiddenness and human philosophy. In A. Green and E. Stump, eds, *Hidden divinity and religious belief*. Cambridge: Cambridge University Press.

Schoenfield, M. 2014. Permission to believe: Why permissivism is true and what it tells us about irrelevant influences on belief. *Noûs* 48: 193–218.

Seyedsayamdost, H. 2015. On gender and philosophical intuition: Failure of replication and other negative results. *Philosophical Psychology* 28: 642–673.

Smith, Q. 1999. The reason the universe exists is that it caused itself to exist. *Philosophy* 74: 579–586.

Street, S. 2006. A Darwinian dilemma for realist theories of value. *Philosophical Studies* 127: 109–166.

Taber, C. S. and M. Lodge. 2006. Motivated skepticism in the evaluation of political beliefs. *American Journal of Political Science* 50: 755–769.

Trakakis, N. 2008. *The end of philosophy of religion*. New York: Continuum.

Trouche, E., E. Sander, and H. Mercier. 2014. Arguments, more than confidence, explain the good performance of reasoning groups. *Journal of Experimental Psychology: General* 143: 1958–1971.

van Inwagen, P. 2010. We're right, they're wrong. In R. Feldman and T. Warfield, eds, *Disagreement*, 10–28. Oxford: Oxford University Press.

White, R. 2005. Epistemic permissiveness. *Philosophical Perspectives* 19: 445–459.

8

An Ecological Theory of Gods' Minds

Benjamin Grant Purzycki and Rita Anne McNamara

Introduction

While the space between the cognitive and evolutionary sciences of religion appears to be shrinking (Watts and Turner 2014), some strains of the cognitive science of religion remain theoretically marginalized from the other evolutionary fields. This is due, in part, to three interpenetrating commitments. First is the view that religion is a functionless or even maladaptive by-product of evolved psychology. Rather than assessed as a hypothesis to be tested, this view was treated as a truism to be repeated (see Alcorta and Sosis 2005; Bulbulia 2008; Sosis 2009; Purzycki et al. 2014). Second, some strains of the field maintain an emphasis on universal cognitive systems as a starting point for inquiry rather than attending to actual religious thought and its variation (cf. Johnson et al. in press). When it does examine explicit thought, it is with the express intent to determine how it conforms or runs counter to universal cognitive architecture or various systems of memory (e.g., Boyer 2001; Barrett 2004; see Appendix). Aside from these commitments, the field pays notably little attention to how religious thought corresponds to features of natural and social environments. As such, there remains a significant—but fillable—void in the field. In this piece, we explore this void and argue that understanding representational models of gods' minds is crucial to understanding religious systems, how they evolve, and how they confront pressing local problems. We present a sketch of a model of the evolution of gods' minds, followed by a scheme by which to test hypotheses about the cross-cultural space that gods' minds appear to occupy.

Evolution of religious systems

According to Sosis and Bulbulia (2011), the field of human behavioral ecology uses "variation in environmental variables to explain variation in human behavior. [Social and natural e]nvironments are vital to the study of adaptive design because traits are only adaptive in relation to a specific environmental context" (343). The behavioral ecology of religion, then, examines how contexts and religious behaviors covary, and whether or not certain behaviors provide fitness benefits in certain contexts. Human behavioral ecologists readily admit to ignoring cognition and are "largely agnostic about the principal mechanisms of adaptive responses." Rather, "behavioral ecologists assume that selection has produced behavior-generating mechanisms that enable organisms to respond optimally—given design constraints and tradeoffs—to environmental conditions" (Sosis and Kiper 2014, 261; see also Smith 2000; Laland and Brown 2011).

Cognitive ecology, however, attends to how variation in social and natural environments explains variation in human cognition (Hutchins 1995, 2010). Real (1993) notes that the goals of cognitive ecology are to "elucidate the underlying psychological and cognitive processes that enter into ecological decision-making, to determine the degree to which these mechanisms are the product of adaptive evolutionary change, and to ascertain the degree to which existing cognitive information-processing schemes constrain potential characterizations of the environment of the organism" (415). An evolutionary cognitive ecological account of religion, then, would ask whether or not, how, and why: (a) contexts and religious *thought* covary, and (b) the content and processes involved in religious thought function to produce adaptive responses to pressures of our social and natural environments (Purzycki et al. 2014).

Anthropologists have long observed that around the world, societies' religious cosmologies often reflect their modes of subsistence and distribution of political or decision-making power (Swanson 1960; Wallace 1966; Peoples and Marlowe 2012). In other words, people often model their religious pantheons on how their societies are structured. However, if religion contributes to the minimization of local problems borne out by subsistence and sociality (Alcorta and Sosis 2005; Purzycki and Sosis 2009, 2010, 2011), then religious beliefs[1] and behaviors should correspond in specific ways that minimize such problems. More specifically, beliefs and practices should predictably change according to shifting socioecological conditions in ways that better manage those conditions, not merely as reflections of other aspects of our societies. The collective action and coordination required to manage these conditions require communication. The content of that communication must motivate others to collaborate, and appeals to gods are excellent candidates for motivating others.

We wager that changing local social and ecological problems that stem from the complex interactions between humans and their natural environments will predict the content of representational models of gods' minds. This model encompasses individual (e.g., *God wants me to do the very best I can do in school*) and collective models of gods' minds (e.g., *The gods don't like it if you disrupt these sacred places*) as discussed below. Religion is a dynamic system; while models of gods' minds might stimulate corresponding behaviors, people are also remarkably adept at rationalizing their behaviors with post-hoc appeals to supernatural agents. This dynamic coupling corresponds to and affects socioecological problems, which in turn feed back to the reinforcement, stabilization, and expression of this coupling (Purzycki and Sosis 2013). Religious cognition largely follows from complex interactions between social and ecological environments not merely as a matter of "learning," "memory," or "transmission," but *as a function of* adaptive behaviors that minimize the costs that local social problems impose. Religion is not a closed system, of course, since other cognitive systems, ontogenetic processes, and external ecological factors influence religious systems and claims of what gods know and care about. Yet, as we discuss below, crucial components of religion often independently converge. Gods' minds are one such component.

Précis of an ecological theory of gods' minds

Boyer (2001, 144) made the observation that when people mention the gods, they tend to be more focused on gods' minds rather than on other features such as appearance. Of course, gods are often visually represented through imagery and statues, but their knowledge and concerns are what make them particularly salient. What makes gods' knowledge and concerns so compelling is the fact that, even though they may be represented as flawed or even powerless in some domains, they are nevertheless more powerful than humans in many respects and therefore worth serious consideration regarding what they want of us (Garcia 2015). People often claim that gods' powers can provide and take away things that natural agents find difficult to achieve on their own (e.g., fertility of crops, rain, and offering supernatural punishment and rewards), and our behavior can alter theirs by virtue of their knowledge and concern.

Gods are often associated with what Boyer calls "socially strategic information," "the subset of all the information ... that activates the mental systems that regulate social interaction" (Boyer 2000, 152, 2002). Defined in this fashion, socially strategic information can be virtually anything given the context. Things as trite as the color of someone's shoes after Labor Day, the careless way one dresses at a job interview, who was having coffee with whom, and so forth are all pieces of information that may activate "the mental systems that regulate social interaction" (Boyer 2000, 152) insofar as someone actually explicitly or implicitly values such things. Gods, however, do not necessarily care about every possible thing that activates social cognitive systems. Rather, specific domains of socially strategic information are more ubiquitously represented as chief among the concerns of specific gods. A more nuanced view that does not rely on the convenience of post-hoc characterization, therefore, needs to account for why this is; which in turn requires a closer look at the *content* of those concerns and the breadth of knowledge people attribute to their gods.

Why gods' minds matter

Psychologically, gods' minds matter to people by virtue of the mentalizing systems that handle them and ongoing developmental psychological research continues to examine how such systems function in relation to beliefs about gods (Heiphetz et al. in press). Drawing on Guthrie's landmark works (1980, 1995), current trends largely maintain the position that making sense of the gods requires making sense of other minds. Presumably, the same evolved systems that allow us to reflect, predict, and act upon the mental states of other people allow us to do the same to the gods (Norenzayan et al. 2012; cf. Reddish et al., 2015). These systems are variously referred to as "theory of mind," "mentalizing," and/or "agency detection" (Premack and Woodruff 1978; Baron-Cohen 1995). Additionally, when primed with agency—regardless of whether or not the agent is a human, ghost, god, or an image of two eyes—people appear to be more likely to follow rules or behave better (Haley and Fessler 2005; Bateson et al. 2006; Waytz et al. 2010; Piazza et al. 2011; Gray et al. 2012; Powell et al. 2012; Nettle et al. 2013). Even though people sometimes do not *explicitly* claim their gods care about human moral behavior, moral cognition still might operate latently, and thus curb antisocial behavior regardless (Purzycki 2011, 2013; Purzycki et al. 2012). Just as pictures of eyes above an office coffee pot might keep people cleaning it more often (Ernest-Jones et al. 2011), indices of gods (e.g., icons, statues, sacred space and place) may prime deeper moral faculties, regardless of gods' culturally specific concerns. We focus here on the explicit content of these faculties' operations.

Socially, religion functions as an organizational strategy, and the gods serve as particular tactics within this strategy (Purzycki and Sosis 2011). In other words, religion can facilitate the reduction of variation in behavior that can contribute to the deterioration of fitness-enhancing relationships. In terms of gods' minds, people appeal to gods in order to manipulate others (e.g., *If you keep doing x, the ancestors will make you ill*; Lee 2003; *if you don't go to the menstrual hut, you'll be struck by lightning because you offended the spirits*; Strassman 1992), to rationalize or justify their actions (e.g., *We give offerings here because the local spirit appreciates it*; Purzycki 2010; *I go to church because God wants me to*), and to motivate behavior (e.g., *I'd better do this so the spirits don't get upset*).

Appeals to gods' minds steer people toward fitness-relevant behaviors; it is striking how closely tied religion and problems of our social and natural environments actually are. More specifically, if we assess representational models of gods, they should, in principle, tell us about much more than how a population thinks about their gods. It should also tell us: (a) how and what people communicate with each other about their gods, (b) why this communication matters to people, (c) the range of local thought that limits the possible content of socially transmitted information, and (d) if and how this information points to shared, local, fitness-relevant problems. If they do, then what gods know and care about should therefore shift accordingly.

What gods know: Relative omniscience

Not all gods are represented as omniscient. Even the omniscience of those popularly held as such (e.g., the Abrahamic god) is often questioned by theologians and devotees alike (Brown 1991; Kapitan 1991; Abbruzzese 1997; see section "Synchronic vs. diachronic beliefs" in Appendix), and the Bible is inconsistent regarding God's omniscience as well. Nevertheless, the breadth of supernatural agents' knowledge (i.e., absolute omniscience being on one end of this scale) has been associated with greater social complexity (Swanson 1960; Wallace 1966; Bering and Johnson 2005; Shariff 2011). Further, supernatural omniscience appears to be cognitively intertwined with gods' concern for morality; the more gods know, the more they appear to care about moral behavior (Purzycki 2013; see below). Yet, whether or not gods around the world are assumed or thought of to be more knowledgeable than normal humans is unknown. Quite likely, the breadth of knowledge people attribute to the gods is a function of the interaction between gods' jurisdiction and the knowledge of the world beyond believers' communities.

What gods care about

Gods appear to be concerned with a limited set of things, including but not necessarily limited to: ritual, moral and immoral behavior, etiquette, and management of resources (Purzycki and Sosis 2011). Unfortunately, however, rich, systematic, quantifiable, and comparable cross-cultural work on beliefs is woefully impoverished. Consequently, we must rely on generations of qualitative ethnographies with low methodological rigor that can lead to suspect generalizations, cross-cultural databases relying on such accounts, and cultural psychologists relying on survey and experimental instruments deployed among western university students (Sears 1986; Henrich et al. 2010) that often assume Christianity as a baseline for generalizations. Drawing from ethnographies and our ongoing research, we class gods' concerns in three general categories: (1) things that people do toward each other, (2) things that people do toward the gods, and (3) things that people do toward nature. In many cases, these domains of gods' concerns are likely to overlap (e.g., the gods do not like it if you hunt on someone else's land). For the sake of organization, we examine each of these categories with this caveat in mind.

Toward people: Morality, virtue, and etiquette

While it is common to assume or claim that religion provides a moral code from which people draw their good behavior (see Zuckerman 2008), cross-cultural evidence suggests that only *some* gods are explicitly associated with behaviors that approximate to the "moral." "Moralistic" gods have been explained ecologically in various ways by virtue of social complexity (e.g., moral gods minimize costs of breaching moral codes; Swanson 1960; Wallace 1966; Botero et al. 2014), resource scarcity (e.g., moral gods foster equitable distribution of resources; Snarey 1996), subsistence and economy (e.g., moral gods build strong bonds for herders who engage in internecine warfare over land; Peoples and Marlowe 2012; cf. Baumard et al. 2015).

At least three related questions should immediately arise when evaluating such studies: (1) whose *morality do such studies refer to?*, (2) *"moral" as compared to what?*, and (3) *what is the source of such data?* (see section "Emic vs. etic"; Johnson in press; Purzycki n.d.). In his assessment of forty-three ethnographies of eighteen different foraging societies, Boehm (2008) distinguishes between moral behaviors and "nonmoral taboos" sanctioned by the gods. Such taboos revolve around "behaviors [that] do not involve being predatory on fellow band members." Among others, these include various violations of food, ritual, birth, and death taboos. His conclusions were that with his sample, foraging societies' deities are far more concerned with "nonmoral taboos" than "morality."

We use "morality" here as explicitly represented, reflective models of interpersonal social behaviors *emically* characterized as "good" or "bad" and/or *etic* characterizations of the moral as normalized behaviors with a cost benefit to others (Turiel 1983; Smetana 2006; Haidt 2008; Rossano 2008; Gray et al. 2012; Greene 2013; Purzycki 2011, 2013). Such *behaviors* might be thought of as distinct from "virtues," which are intrinsic *qualities* of individuals that may or may not have obvious social ramifications or behavioral corollaries. So, killing someone is a moral affair whereas being lazy might be more a matter of virtue. Both of these domains are obvious candidates for universal models of what constitutes a "good" or "bad" person; such qualities and behaviors clearly contribute to human sociality.

Supernatural agents believed to be monitors and/or punishers of violations of morality, and virtue may be particularly effective in promoting cooperation and suppressing antisocial behavior (Johnson and Krüger 2004; Johnson 2005; Shariff and Norenzayan 2007, 2011; Atkinson and Bourrat 2011; Schloss and Murray 2011). The effects of supernatural monitoring and punishment cognition may be especially important in situations where other institutionalized, secular forms of social control are ineffective or absent (Norenzayan 2013).[2] And again, such models should correspond to highly complex societies with greater anonymity and affordances to engage in "immoral" behavior.

Take, for instance, religious advocacy of violence. Matthews and colleagues (2013) find that religious advocacy of violence among Anabaptists corresponds less to the transmission of theological positions than to local contexts. Incorporating this finding into the model requires inclusion of the effects of out-group members' behaviors, rhetoric, and so forth. These feed into socioecological problems by way of the complex relationship between subsistence, natural environment, and features of populations. Beliefs about whether gods care about how we treat other people are also often highly dependent upon who might be the target of one's actions. Socio-environmental threats that societies face, including disease burden and threat of intergroup violence, may be particularly important in influencing whom God thinks believers should be nice to. Believing that the Abrahamic God condones violence may correspond to aggressive responses

(Bushman et al. 2007), while belief that He is powerful and in control can suppress both individual action in punishing others as well as individual support for worldly punishments through secular institutions (Laurin et al. 2012).

Etiquette—arbitrary behavioral conventions—often conveys social influence and control and functions as behavioral badges and reliable indicators of costs involved in social inclusion (Strassman 1992; Sosis and Bressler 2003; Sosis 2006). Appeals to the gods' concerns of etiquette may function as a means to maintain "appropriate" social conduct and reap benefits to which those behaviors may correspond. In Fiji, for instance, local deified ancestor spirits care when one wears hats in the village or shouts in the village at night. Though these may seem trivial, these kinds of etiquette violations convey the violator's disrespect and contempt for the traditional kin-based hierarchy. Such violations are met with disrespect and anger from the living, which has its own dimension of supernatural danger, as well as more direct supernatural dangers from ancestor spirits' retributive curses of bad luck and illness (Katz 1999; Gervais 2013). These proscriptions against etiquette violations corroborate the sacredness of interpersonal obligations within traditional Fijian social structures that are essential to successful subsistence in a sometimes risky environment (Anae 2010; Gervais 2013; McNamara et al. in press). Supernatural sanctions that compel menstruating women among the Dogon of Mali to temporarily reside in public menstrual huts similarly cue social information that may support traditional societal structures. Premenopausal women's periodic residence in these menstrual huts reliably signals their current reproductive status as well as the strong influence that men (and the gods) exert on married women's lives. This in turn leads to reduced cuckoldry and increased paternity certainty (Strassmann 1992; Strassmann et al. 2012).

Toward the gods: Belief and ritual

While "belief" (i.e., propositional faith) is undoubtedly important in some Abrahamic and Buddhist theologies, and often assumed to be a central facet of religion, it is not central to most traditions (Fernandez 1965; Sosis and Kiper 2014). Emphasis on "belief" in gods is likely increasingly important in large-scale societies with stronger, worldly social controls, nonethnic, universalizing religious traditions, and those traditions under threat of dissolution (Purzycki and Sosis 2011). Relatively secular, socially complex societies provide nonreligious and non-kin sources of ordering social interactions, giving individuals more freedom to act as autonomous agents. This individual autonomy can break down the communicative force of public ritualized behavior in favor of showing other signs of proper internal orientations to the divine (Armstrong 2006; Taylor 2007; Cavanaugh 2011; Lanman 2012). In these cases, declaring one's belief or faith becomes essential. Thought may become increasingly equated with action and indicative of what it means to be appropriately "religious" (Cohen and Rozin 2001; Cohen et al. 2003).

Ritual, however, is central to many—if not most—gods' concerns. While rituals might correspond to various kinds of memory stores (Whitehouse 2004), and while we might expend important, locally available resources to pay our respects (e.g., gods prefer sacrifices of pigs rather than grass), a more profound possibility is that rituals (and the gods to whom these rituals are devoted) might also evolve in specific ways that minimize the deleterious effects of certain social and ecological challenges (Purzycki and Sosis 2009; Atkinson and Whitehouse 2011). Of the former, this is indicated primarily by variation in ritual costs, timing, and location.

Paying ritual costs publicly conveys a reliable sense of otherwise inaccessible private mental states (Rappaport 1999). By engaging in ritual, people convey how committed they are to both the gods and social partners. These costs translate to more reliable and sustained cooperation that is facilitated by trust (Sosis and Bressler 2003; Sosis and Ruffle 2003; Sosis 2005; Soler 2012; Purzycki and Arakchaa 2013; Xygalatas et al. 2013). It follows that local variations in threats to bonds should correspond to variation in the costs and spatiotemporal distributions of rituals.

When groups face threats that make disbanding from the group increasingly tempting (e.g., warfare), rituals increase in risk and cost (Sosis et al. 2007). Places with high warfare frequency engage in genital mutilation, tooth extraction, scarification, and other tests of mettle more often than more peaceful societies. Ritual costs do not always have to be extreme, but even relatively cost-free rituals communicate to observers. Rituals also mark the passage of time and revolve around important life stages cross-culturally (Reynolds and Tanner 1995; Shaver and Sosis 2014; Shaver, in press). They also are timed when resources require management (Rappaport 1984; Lansing 1987, 2007; Lansing and Kremer 1993). And, ritual places are often spatially located at territorial borders (Jordan 2003; Purzycki 2010). Notably, localization of spirits via ritual places corresponds to the breadth of moral knowledge attributed to spirits (Purzycki 2013).

Cross-culturally, many herding and hunting groups appear to have independently evolved a ritual system where spirits are associated with travel and territory (Sierksma 1963). For example, the Hopi traditionally viewed one particular deity's role "as owner and guardian spirit of the earth … [who] assumes protectorship of those who travel on his land. To be granted a good journey the Hopi traveler formerly would leave a votive offering at one of the god's shrines" (Malotki and Lomatuway'ma 1987, 84). As "owner of the land," this spirit gave humans his permission to live on it (Waters 1963, 22). Compare this to many Inner Asian groups, where local "spirit-masters" are also thought of as "owners" of the earth. During travelling, people make ritualized offerings to these spirits at cairns located on territorial borders. They also do this for safe passage.

Nonlocal visitors to Ainu villages required permission from local leaders and spirits to hunt and fish. Permission was granted through ritual performance, and trespassers were brought before leaders and required to apologize to the spirits for their offense (Watanabe 1972). Fisher-horticulturalists in Fiji also require ritual offerings for local spirits' approval whenever anyone enters or exits a village for any extended time. The ancestors are the traditional protectors of the *vanua* ("people" and "land"), or the inseparable interdependence between Fijian clans and their ancestral lands. Seeking the ancestor's blessing in these entrance and exit rituals corroborates spacio-social connections by acknowledging the inherent danger in entering another clan's territory or distancing oneself from the kin network one is obliged to support. The Khanty of western Siberia are a hunter-gatherer-fisher population living in small communities. Each community has its own traditional hunting grounds. While they seasonally move around in these territories in order to exploit particular resources, they always return to riverside base camps and engage in collective religious rites. If an individual enters someone else's community or hunting territory, one must make offerings to the local spirits (Jordan 2003, 146). Given the social effects of ritual and the priming of supernatural agency, these ritual patterns that reliably convey submission to local expectations may minimize the costs involved in conflict over territory and resources (see Sosis 2011; Johnson and Toft 2014).

Toward nature: Ecology and resource maintenance

Gods also often care about how we treat natural resources. Indigenous intellectuals consistently emphasize the sacred in discussions of humanity's relationship to nature (LaDuke 1999; Battiste and Henderson 2000; Deloria, Jr. 2003), and the burgeoning field of "eco-theology" is refashioning how people in the Abrahamic traditions talk about their relationship to God and the preservation of the natural world (White 1967; Sponsel 2014). While cultural ecological works detail evidence that religion can contribute to ecological knowledge in a variety of ways (Klubnikin et al. 2000; Atran et al. 2002; Berkes 2012), too few examine whether or not religious postulates reflect adaptive ecological behavior.

Among the hunters and herders in the Tyva Republic of southern Siberia, local "spirit-masters" are primarily angered when people sully, pollute, and overexploit natural resources that "belong" to them (Purzycki 2011). Additionally, some areas are off-limits to hunting because they are considered to be spirits' territories (see Donahoe 2003, 111–146). This may indeed function to preserve game. Again, however, such beliefs do not necessarily mean that people actually engage in these prescribed behaviors. Indeed, there is considerable skepticism toward the view of the "native environmentalist" (Smith and Wishnie 2000; Hames 2007). However, there is some evidence that religiously rationalized and motivated resource management can provide fitness benefits for constituents.

Rappaport (1984) famously investigated the relationship between intergroup conflict, religion, and pig-rearing practices among the Tsembaga in Papua New Guinea. As pigs eat the same resources as people, having too many creates strain on gardens and contributes to environmental degradation that result in conflict over territories. Pigs are slaughtered predominantly during spiritual devotions timed during stressful territorial conflicts. More recent studies show that religiously prescribed behaviors show significant effects on caloric returns in hunting practices that contribute to biodiversity (Bliege Bird et al. 2013; Codding et al. 2014). According to Bliege Bird and colleagues (2013), the Australian Martu "believe that if they do not continue to reenact the Jukurrpa [sacred law] through emulating the creative forces of the ancestral beings across the landscape—hunting, collecting, burning and caring for family—those plants and animals that depend on their actions will cease to exist" (2).

Conclusion

Representational models of gods' minds are the content of human communication with respect to the things about which the gods care. They may point to the behaviors that alter the chances that one reproduces, but they are not the behaviors (or absence of behaviors) themselves which may or may not provide fitness benefits. As such, getting a sense of gods' minds is a useful research strategy to further investigate specific case studies and contribute to the discussion of whether or not and how religion evolves because it confers fitness advantages. Behavior and its effects that accompany beliefs are crucial for any satisfactory account of the adaptiveness of religiosity.

Considerable work suggests that while not unidirectional, humans' interactions with their natural and social environments predict the kinds of beliefs and rituals people have. If this is the case, we predict that components of religious beliefs and cognition will vary as a function of these complex interactions. Representational models of gods' minds are one such component. Our outline of how to approach the study of beliefs about gods' thoughts, concerns, and abilities can give a meaningful structure to decode the dizzying array of variation in religious forms that are often still treated as noise by much existing scientific work on religion. This approach requires extensive and deliberate documentation of how people talk about their gods, what they agree upon, and, equally important, what they do not say or agree about when it comes to these supernatural agents. We posit further that it is necessary but not sufficient to include the cognitive and broad, human universal aspects of psychology and cultural transmission biases. Rather, a thorough approach to understanding the form and function of gods' minds includes the socioecological challenges that societies face. Arguably, this is not merely an obvious candidate for further research. Rather, it is crucial and significant for cognitive, cultural, and evolutionary sciences of religion.

Acknowledgments

We thank Nicholas Chan and Tiffany Lai for their discussions and hard work on these issues, to the anonymous reviewers and Ryan Nichols who provided very helpful feedback on an earlier draft of this chapter, and to Helen De Cruz and Ryan Nichols for the invitation to contribute. The Cultural Evolution of Religion Research Consortium (CERC) supported us financially during the preparation of this manuscript. CERC is financially supported by grants from SSHRC and the John Templeton Foundation.

Appendix

Curiously, the methodological and theoretical tools of cognitive anthropology (D'Andrade 1995; Kronenfeld et al. 2011) have yet to fully penetrate the cognitive science of religion. This is likely due to the latter's general commitment to understanding the biological bases for religious thought. Cognitive anthropology has a lot to contribute on this front given its emphasis on the structure and content of explicit and implicit thought. In this appendix, we discuss the focal kinds of beliefs discussed in the literature to highlight those which the cognitive science of religion largely attends and those which cognitive anthropology would examine.

Emic versus *etic*

Consider the case of gods' moral concern (section "Toward people: Morality, virtue, and etiquette"). Despite the volume and importance of these studies seeking to address the social and ecological contexts that give rise to moralistic gods, not a single one of them draws from data systematically collected from people about what their gods care about. Moreover, how representative this moralism is among any given sample is unknown. Additionally, whether or not the determination of gods' "moralism" is drawn from *etic* or *emic* models of morality remains unclear. Indeed, the Standard Cross-Cultural Sample, a database often used for such studies, defines "moralistic High God" as: "a spiritual being who is believed to have created all reality and/or to be its ultimate governor, even though his/her sole act was to create other spirits who, in turn, created or control the natural world" that people view as "Present, active, and specifically supportive of human morality." One interpretation of this is that a specific god *cares about* human morality. If so, the obvious question is: *moral to whom?*

A classical distinction in cultural anthropology is that of the *etic* and *emic* (Pike 1967; Harris 1976; Headland et al. 1990). Crudely put, *emic* data is that which participants (i.e., informants, subjects, interviewees, etc.) offer. So, when we ask people to tell us what the gods care about, they give us *emic* data on their views of the contents of gods' minds. *Etic* information, however, is external to the participant. Theories, typologies, definitions, interpretations, re-characterizations, and so forth are *etic* constructs; our informants are not likely to appeal to such characterizations when they answer our questions, but the data they do provide are *emic*.[3] If *emic* representations are representations of participants' worlds, then *etic* representations are representations of participants' representations. If we are interested in variation in religious beliefs, collecting *emic* data is crucial. This raises another question related to gods' moralism: *what do all of the gods that aren't "specifically supportive of human morality" care about?* One simple way to satisfactorily answer these questions is to ask people what gods care about, craft representative models of these attributed concerns, examine these models' cross-cultural variability, and determine whether or not this variation corresponds to variation in local problems. To do this requires examining accessible, or "reflective" beliefs, not merely the coding of text materials that reflect the methodological limitations and biases of its authors and subsequent coders.

Reflective versus intuitive beliefs

Some (Sperber 1997; Barrett 2004; Slone 2004) emphasize the distinction between "intuitive" and "reflective" beliefs, which often correspond to "naturalness" and "unnaturalness" of religion (Barrett 2000; Bloom 2007; Geertz and Markússon 2010; McCauley 2011). As a result, much effort has focused on identifying how religious belief is effortless and often beyond conscious control. For example, such intuitive, heuristic models of gods' minds should be anthropomorphized to suit evolved social cognitive mechanisms used to understand human agents. This focus on intuitive religious belief also assumes that any representation of a supernatural agent that does not fit the effortless, heuristic model of gods' minds will necessarily only arise due to effortful, reflective thinking (e.g., Barrett and Lanman 2008; Bering 2010). This model may well represent how most people experience gods' minds—either due to lack of ability or interest in thinking more reflectively about gods, or to the fact that all humans use such unreflective, implicit processing for a majority of our basic cognitive functions. Therefore, the majority of psychological work thus far has focused on the unconscious, heuristic model of the Abrahamic god's mind rather than the reflective model.

By way of illustration, Figure 8.1 represents a hypothetical belief set of one aspect of God's concerns. Assume that someone has collected naturalistic data from a sample regarding what the Abrahamic God dislikes (i.e., "Please list 10 things that make God angry"). Numerical values on the model indicate proportionate frequency or salience of each item for the sample as a whole (see Quinlan 2005), and the eight nodes represent the eight most frequent or salient items listed by a sample. This in turn functions as a predictive model of what any single individual within a sample will say, in what order, what kinds of information gets engaged when thinking about God, and perhaps indicative of what kind of deeper cognitive systems get recruited during this event. Such a model allows us to directly compare—both quantitatively and qualitatively—representational models of gods' concerns across individuals, traditions, contexts, time, and other dimensions we might have some theoretical reason to manipulate. This kind of belief set is our dependent

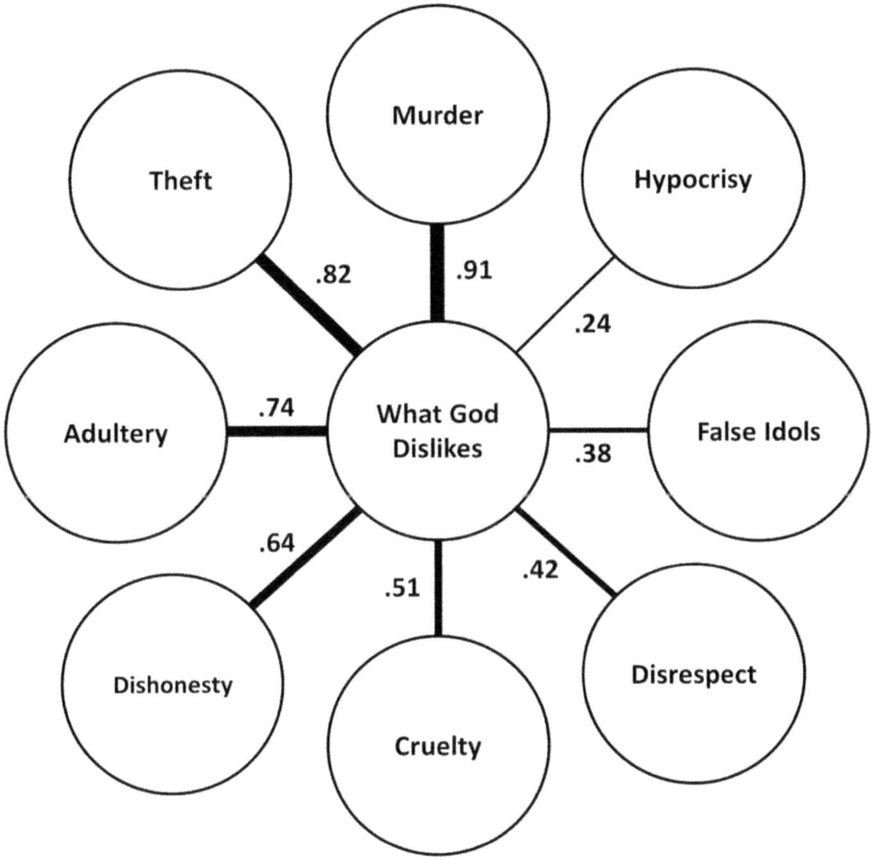

Figure 8.1 Hypothetical representational model of what the Abrahamic God dislikes.

variable (see Purzycki n.d.). Of course, such belief sets change through time, a point to which we now turn.

Synchronic versus diachronic beliefs

Consider the popular notion of "theological correctness" (Barrett 1998, 1999; Slone 2004). While one might say Christians believe that "God knows everything," evidence suggests that our minds do not consistently operate as though they believe this, and we often say things that are inconsistent with this (Purzycki et al. 2012; Purzycki 2013; Lane et al. 2014). What determines this "correctness" is a matter of authorities and experts (including holy books); our minds produce theologically "incorrect" statements by virtue of other cognitive systems' default, intuitive operations. However, authoritative sources are *not* consistent. In other words, we are *supposed* to say that "God knows everything," but even the Bible is not consistent about this point, let alone theologians. Additionally, in many cases, there is no such central doctrine, no central authority, and no written text by which to appeal in order to determine what people are "supposed" to believe.

So, such a notion presupposes—rather than tests—that there is some model out there to which people are expected to conform. As such, we emphasize the importance of "cultural consistency" as a matter of representational consensus rather than "correctness." Cultural or representational consensus models are predictive models of what individuals are likely to or should believe (i.e., ought to believe) given the variance of the sample (Romney et al. 1986; Oravecz et al. 2014). Thus, there is a method that accounts for: (a) individual models, (b) shared models, (c) variation across sharedness for both individuals and groups, and (d) provides scores for how close individuals approximate to those shared models (see Purzycki 2013).

Nevertheless, we have to come to terms with the variety of ways people appeal to gods, the relative stability of the content of such appeals, and the methodological tools required to appropriately characterize any given sample's belief sets, however internally inconsistent they may be. Such beliefs and their stability fluctuate through time and space synchronically and diachronically (Purzycki and McNamara in press). Synchronic reflective beliefs are religious appeals that have a relatively short life span and are highly dependent on situational contexts. So, one might claim that God helped them score a winning basket, but ideas like "God cares about basketball," or "God helps specific people shoot baskets," are neither widespread nor sustained outside of such contexts and times. However, if one were interviewing recovering drug addicts about what God cares about, "God wants me to conquer my addiction," might be a consistent feature of their models and indicative of a sample's shared, persistent problems. Diachronic reflective beliefs are those that have staying power; they are more likely to be shared and are not simply situational postulates. Any ecologically oriented theory of gods' minds predicts that sustained models of what the gods care about will correspond to sustained social and ecological problems, and short-term ecological pressures may correspond to synchronic beliefs about gods' concerns. Therefore, in order to properly situate beliefs in local contexts, we must be careful about the distinction between *when* and *where* people express such beliefs.

Notes

1 We refer to "belief" in two general senses. In some cases, we simply use it to refer to any mental representation. However, we primarily use it to refer to explicit and reflective representational models of spiritual content. In other words, when we say that gods care about X or Y, we are saying that people P's consensus models of gods' minds consist—or should consist—of X or Y. Our present argument is that X or Y should in some way point people to social and ecological challenges to fitness in specific contexts. Note that this argument neither requires nor assumes that individuals actually believe (i.e., hold something to be true) the content of their religious postulates. Rather, we refer to representational models. What we wish to examine is whether or not those representational models (specifically gods' minds) qualitatively correspond to behaviors that moderate challenges to fitness and if so, how and why. Empirically examining such a question requires examining actual belief sets, something to which the cognitive science of religion has largely not attended. See the Appendix for further discussion.

2 This is not to say that societies with weak secular institutions for maintaining cooperative social norms will necessarily develop beliefs in supernatural agents that observe and punish norm violators. Rather, once these beliefs are present, they should be very successful in supporting maintenance of prosocial behavior.

3 *Etically*, it is easy for us to characterize data however we wish. So, one could easily characterize such data as obviously "moral" or "socially strategic information" (see below) in a post-hoc fashion.

Bibliography

Abbruzzese, J. E. 1997. The coherence of omniscience: A defense. *International Journal for Philosophy of Religion* 41: 25–34.

Alcorta, C. S. and R. Sosis. 2005. Ritual, emotion, and sacred symbols: The evolution of religion as an adaptive complex. *Human Nature* 16(4): 323–359.

Anae, M. 2010. Teu Le Va: Toward a native anthropology. *Pacific Studies* 33(2/3): 222–240.

Armstrong, K. 2006. *The great transformation: The beginning of our religious traditions.* New York: Random House.

Atkinson, Q. D. and P. Bourrat. 2011. Beliefs about God, the afterlife and morality support the role of supernatural policing in human cooperation. *Evolution and Human Behavior* 32(1): 41–49.

Atkinson, Q. D. and H. Whitehouse. 2011. The cultural morphospace of ritual form: Examining modes of religiosity cross-culturally. *Evolution and Human Behavior* 32: 50–62.

Atran, S., D. Medin, N. Ross, E. Lynch, V. Vapnarsky, E. U. Ek', J. Coley, C. Timura, and M. Baran. 2002. Folkecology, cultural epidemiology, and the spirit of the commons: A garden experiment in the Maya lowlands, 1991–2001. *Current Anthropology* 43: 421–450.

Baron-Cohen, S. 1995. *Mindblindness: An essay on autism and theory of mind*. Cambridge: The MIT Press.

Barrett, J. L. 1998. Cognitive constraints on Hindu concepts of the divine. *Journal for the Scientific Study of Religion* 37: 608–619.

Barrett, J. L. 1999. Theological correctness: Cognitive constraint and the study of religion. *Method and Theory in the Study of Religion* 11: 325–339.

Barrett, J. L. 2000. Exploring the natural foundations of religion. *Trends in Cognitive Sciences* 4: 29–34.

Barrett, J. L. 2004. *Why would anyone believe in God*? Walnut Creek, CA: AltaMira Press.

Barrett, J. L. and J. A. Lanman. 2008. The science of religious beliefs. *Religion* 38: 109–124.

Bateson, M., D. Nettle, and G. Roberts. 2006. Cues of being watched enhance cooperation in a real-world setting. *Biology Letters* 2: 412–414.

Battiste, M. and J. Y. Henderson. 2000. *Protecting indigenous knowledge and heritage: A global challenge*. Saskatoon: Purich Publishing Ltd.

Baumard, N., A. Hyafil, I. Morris, and P. Boyer. 2015. Increased affluence explains the emergence of ascetic wisdoms and moralizing religions. *Current Biology* 25(1): 10–15.

Bering, J. M. 2010. Atheism is only skin deep: Geertz and Markusson rely mistakenly on sociodemographic data as meaningful indicators of underlying cognition. *Religion* 40(3): 166–168.

Bering, J. M. and D. D. P. Johnson. 2005. "O Lord … You perceive my thoughts from afar": Recursiveness and the evolution of supernatural agency. *Journal of Cognition and Culture* 5: 118–142.

Berkes, F. 2012. *Sacred ecology*. New York: Routledge.

Bliege Bird, R. B., N. Tayor, B. F. Codding, and D. W. Bird. 2013. Niche construction and dreaming logic: Aboriginal patch mosaic burning and Varanid Lizards (*Varanus gouldii*) in Australia. *Proceedings of the Royal Society B* 280: 20132297.

Bloom, P. 2007. Religion is natural. *Developmental Science* 10(1): 147–151.

Boehm, C. 2008. A biocultural evolutionary exploration of supernatural sanctioning. In J. Bulbulia, R. Sosis, E. Harris, R. Genet, and K. Wyman, eds, *Evolution of religion: Studies, theories, and critiques*, 143–152. Santa Margarita, CA: Collins Foundation Press.

Botero, C. A., B. Gardner, K. R. Kirby, J. Bulbulia, M. C. Gavin, and R. D. Gray. 2014. The ecology of religious beliefs. *Proceedings of the National Academy of Sciences* 111(47): 16784–16789.

Boyer, P. 2000. Functional origins of religious concepts: Ontological and strategic selection in evolved minds. *The Journal of the Royal Anthropological Institute* 6: 195–214.

Boyer, P. 2001. *Religion explained: The evolutionary origins of religious thought*. New York: Basic Books.

Boyer, P. 2002. Why do gods and spirits matter at all? In I. Pyysiäinen and A. Veikko, eds, *Current approaches in the cognitive science of religion*, 68–92. New York: Continuum.

Brown, R. F. 1991. Divine omniscience, immutability, aseity and human free will. *Religious Studies* 27: 285–295.

Bulbulia, J. 2008. Meme infection or religious niche construction?: An adaptationist alternative to the cultural maladaptationist hypothesis. *Method and Theory in the Study of Religion* 20: 67–107.

Bushman, B. J., R. D. Ridge, E. Das, C. W. Key, and G. L. Busath. 2007. When God sanctions killing: Effect of scriptural violence on aggression. *Psychological Science* 18(3): 204–207.

Cavanaugh, W. T. 2011. *Migrations of the holy: God, state, and the political meaning of the church*. Grand Rapids, MI: Wm. B. Eerdmans Publishing Company.

Codding, B. F., R. B. Bird, P. G. Kauhanen, and D. W. Bird. 2014. Conservation or co-evolution? Intermediate levels of aboriginal burning and hunting have positive effects on kangaroo populations in western Australia. *Human Ecology* 42: 659–669.

Cohen, A. B. and P. Rozin. 2001. Religion and the morality of mentality. *Journal of Personality and Social Psychology* 81(4): 697–710.

Cohen, A. B., J. I. Siegel, and P. Rozin. 2003. Faith versus practice: Different bases for religiosity judgments by Jews and Protestants. *European Journal of Social Psychology* 33: 287–295.

D'Andrade, R. G. 1995. *The development of cognitive anthropology*. New York: Cambridge University Press.

Deloria, Jr., V. 2003. *God is red: A native view of religion*. Golden, CO: Fulcrum Publishing.

Donahoe, B. R. 2003. *A line in the Sayans: History and divergent perceptions of property among the Tozhu and Tofa of south Siberia*. Unpublished Dissertation. Bloomington: University of Indiana.

Ernest-Jones, M., D. Nettle, and M. Bateson. 2011. Effects of eye images on everyday cooperative behavior: A field experiment. *Evolution and Human Behavior* 32: 172–178.

Fernandez, J. W. 1965. Symbolic consensus in a Fang reformative cult. *American Anthropologist* 67(4): 902–929.

Garcia, H. A. 2015. *Alpha God: The psychology of religious violence and oppression*. New York: Prometheus Books.

Geertz, A. W. and G. I. Markússon. 2010. Religion is natural, atheism is not: On why everybody is both right and wrong. *Religion* 40(3): 152–165.

Gervais, M. M. 2013. *Structures of sentiment: Mapping the affective bases of social relationships in Yasawa, Fiji*. Los Angeles, CA: University of California Los Angeles. Retrieved from http://gradworks.umi.com/35/94/3594741.html

Gray, K., L. Young, and A. Waytz. 2012. Mind perception is the essence of morality. *Psychological Inquiry* 23: 101–124.

Greene, J. 2013. *Moral tribes: Emotion, reason, and the gap between us and them*. New York: Penguin.

Guthrie, S. E. 1980. A cognitive theory of religion. *Current Anthropology* 21: 181–203.

Guthrie, S. E. 1995. *Faces in the clouds: A new theory of religion*. New York: Oxford University Press.

Haidt, J. 2008. Morality. *Perspectives on Psychological Science* 3: 65–72.

Haley, K. J. and D. M. T. Fessler. 2005. Nobody's watching?: Subtle cues affect generosity in an anonymous economic game. *Evolution and Human Behavior* 26: 245–256.

Hames, R. 2007. The ecologically noble savage debate. *Annual Review of Anthropology* 36(1): 177–190.

Harris, M. 1976. History and significance of the emic/etic distinction. *Annual Review of Anthropology* 5: 329–350.

Headland, T. N., K. Pike, and M. Harris. 1990. *Emics and etics: The insider/outsider debate*. Newbury Park, CA: SAGE Publications.

Heiphetz, L., J. D. Lane, A. Waytz, and L. L. Young. in press. How children and adults represent God's mind. *Cognitive Science*.

Henrich, J., S. J. Heine, and A. Norenzayan. 2010. The weirdest people in the world? *Behavioral and Brain Sciences* 33: 61–83.

Hutchins, E. 1995. *Cognition in the wild*. Cambridge, MA: MIT Press.

Hutchins, E. 2010. Cognitive ecology. *Topics in Cognitive Science* 2(4): 705–715.

Johnson, D. D. P. 2005. God's punishment and public goods: A test of the supernatural punishment hypothesis in 186 world cultures. *Human Nature* 16(4): 410–446.

Johnson, D. D. P. in press. Big gods, small wonder: Supernatural punishment strikes back. *Religion, Brain and Behavior*.

Johnson, D. D. P. and O. Krüger. 2004. The good of wrath: Supernatural punishment and the evolution of cooperation. *Political Theology* 5(2): 159–176.

Johnson, D. D. P. and M. D. Toft. 2014. Grounds for war: The evolution of territorial conflict. *International Security* 38(3): 7–38.

Johnson, K. A., J. L. Yexin, and A. B. Cohen. 2015. Fundamental social motives and the varieties of religious experience. *Religion, Brain and Behavior* 5(3): 197–231.

Jordan, P. 2003. *Material culture and sacred landscape: The anthropology of the Siberian Khanty*. Walnut Creek, CA: Altamira Press.

Kapitan, T. 1991. Agency and omniscience. *Religious Studies* 27: 105–120.

Katz, R. 1999. *The straight path of the spirit: Ancestral wisdom and healing traditions in Fiji*. Rockester: Park Street Press.

Klubnikin, K., C. Annett, M. Cherkasova, M. Shishin, and I. Fotieva. 2000. The sacred and the scientific: Traditional ecological knowledge in Siberian river conservation. *Ecological Applications* 10: 1296–1306.

Kronenfeld, D. B., G. Bennardo, V. C. de Munck, and M. D. Fischer. 2011. *A companion to cognitive anthropology*. Malden, MA: Wiley-Blackwell.

LaDuke, W. 1999. *All our relations: Native struggles for land and life*. Cambridge: South End Press.

Laland, K. N. and G. R. Brown. 2011. *Sense and nonsense: Evolutionary perspectives on human behaviour*. Oxford: Oxford University Press.

Lane, J. D., H. M. Wellman, and E. M. Evans. 2014. Approaching an understanding of omniscience from the preschool years to early adulthood. *Developmental Psychology* 50: 2380–2392.

Lanman, J. A. 2012. The importance of religious displays for belief acquisition and secularization. *Journal of Contemporary Religion* 27(1): 49–65.

Lansing, J. S. 1987. Balinese "water temples" and the management of irrigation. *American Anthropologist* 89: 326–341.

Lansing, J. S. 2007. *Priests and programmers: Technologies of power in the engineered landscape of Bali*. Princeton, NJ: Princeton University Press.

Lansing, J. S. and J. N. Kremer. 1993. Emergent properties of Balinese water temple networks: Coadaptation on a rugged fitness landscape. *American Anthropologist* 95: 97–114.

Laurin, K. et al. 2012. Outsourcing punishment to God: Beliefs in divine control reduce earthly punishment. *Proceedings of the Royal Society B: Biological Sciences* 279(1741): 3272–3281.

Lee, R. B. 2003. *The Dobe Ju/'Hoansi*. Belmont, CA: Wadsworth Publishing.

Malotki, E. and M. Lomatuway'ma. 1987. *Maasaw: Profile of a Hopi god*. Lincoln: University of Nebraska Press.

Matthews, L. J., J. Edmonds, W. J. Wildman, and C. L. Nunn. 2013. Cultural inheritance or cultural diffusion of religious violence? A quantitative case study of the Radical Reformation. *Religion, Brain and Behavior* 3: 3–15.

McCauley, R. N. 2011. *Why religion is natural and science is not*. Oxford: Oxford University Press.

McNamara, R. A., A. Norenzayan, and J. Henrich. in press. Supernatural punishment, in-group biases, and material insecurity: Experiments and ethnography from Yasawa, Fiji. *Religion, Brain and Behavior*.

Nettle, D., Z. Harper, A. Kidson, R. Stone, I. S. Penton-Voak, and M. Bateson. 2013. The watching eyes effect in the dictator game: It's not how much you give, it's being seen to give something. *Evolution and Human Behavior* 34: 35–40.

Norenzayan, A. 2013. *Big gods: How religion transformed cooperation and conflict*. Princeton, NJ: Princeton University Press.

Norenzayan, A., W. M. Gervais, and K. H. Trzesniewski. 2012. Mentalizing deficits constrain belief in a personal god. *PLoS ONE* 7: e36880.

Oravecz, Z., J. Vandekerckhove, and W. H. Batchelder. 2014. Bayesian cultural consensus theory. *Field Methods* 26(3): 207–222.

Peoples, H. C. and F. W. Marlowe. 2012. Subsistence and the evolution of religion. *Human Nature* 23(3): 253–269.

Piazza, J., J. M. Bering, and G. Ingram. 2011. "Princess Alice is watching you": Children's belief in an invisible person inhibits cheating. *Journal of Experimental Child Psychology* 109: 311–320.

Pike, K. L. 1967. *Language in relation to a unified theory of the structure of human behavior*. The Hague: Mouton.

Powell, K. L., G. Roberts, and D. Nettle. 2012. Eye images increase charitable donations: Evidence from an opportunistic field experiment in a supermarket. *Ethology* 118: 1096–1101.

Premack, D. and G. Woodruff. 1978. Does the chimpanzee have a theory of mind? *Behavioral and Brain Sciences* 1(4): 515–526.

Purzycki, B. G. 2010. Spirit masters, ritual cairns, and the adaptive religious system in Tyva. *Sibirica* 9: 21–47.

Purzycki, B. G. 2011. Tyvan *cher eezi* and the socioecological constraints of supernatural agents' minds. *Religion, Brain and Behavior* 1: 31–45.

Purzycki, B. G. 2013. The minds of gods: A comparative study of supernatural agency. *Cognition* 129: 163–179.

Purzycki, B. G. n.d. Representational models of gods' minds in the Tyva Republic.

Purzycki, B. G. and T. Arakchaa. 2013. Ritual behavior and trust in the Tyva Republic. *Current Anthropology* 54: 381–388.

Purzycki, B. G., D. N. Finkel, J. Shaver, N. Wales, A. B. Cohen, and R. Sosis. 2012. What does God know? Supernatural agents' access to socially strategic and non-strategic information. *Cognitive Science* 36: 846–869.

Purzycki, B. G., O. S. Haque, and R. Sosis. 2014. Extending evolutionary accounts of religion beyond the mind: Religions as adaptive systems. In F. Watts and L. P. Turner, eds, *Evolution*, religion, and cognitive science: Critical and constructive essays, 74–91. Oxford: Oxford University Press.

Purzycki, B. G. and R. A. McNamara. 2015. Ecology, consensus, and variation: Issues with time and persistence in religious systems. *Religion, Brain and Behavior*, 5(3): 250–253.

Purzycki, B. G. and R. Sosis. 2009. The religious system as adaptive: Cognitive flexibility, public displays, and acceptance. In E. Voland and W. Schiefenhövel, eds, *The biological evolution of religious mind and behavior*, 243–256. New York: Springer.

Purzycki, B. G. and R. Sosis. 2010. Religious concepts as necessary components of the adaptive religious system. In U. Frey, ed., *The nature of God: Evolution and religion*, 37–59. Marburg: Tectum Verlag.

Purzycki, B. G. and R. Sosis. 2011. Our gods: Variation in supernatural minds. In U. J. Frey, C. Störmer, and K. P. Willführ, eds, *Essential building blocks of human nature*, 77–93. New York: Springer-Verlag.

Purzycki, B. G. and R. Sosis. 2013. The extended religious phenotype and the adaptive coupling of ritual and belief. *Israel Journal for Ecology and Evolution* 59(2): 99–108.

Quinlan, M. 2005. Considerations for collecting freelists in the field: Examples from ethnobotany. *Field Methods* 17(3): 219–234.

Rappaport, R. A. 1984. *Pigs for the ancestors: Ritual in the ecology of a New Guinea people*. Long Grove, IL: Waveland Press Inc.

Rappaport, R. A. 1999. *Ritual and religion in the making of humanity*. Cambridge: Cambridge University Press.

Real, L. A. 1993. Toward a cognitive ecology. *Trends in Ecology and Evolution* 8(11): 413–417.

Reddish, P., P. Tok, and R. Kundt. 2015. Religious cognition and behaviour in autism: The role of mentalizing. *The International Journal for the Psychology of Religion* 1–36.

Reynolds, V. and R. Tanner. 1995. *The social ecology of religion*. New York: Oxford University Press.

Romney, A. K., S. C. Weller, and W. H. Batchelder. 1986. Culture as consensus: A theory of culture and informant accuracy. *American Anthropologist* 88(2): 313–338.

Rossano, M. J. 2008. The moral faculty: Does religion promote "moral expertise"? *International Journal for the Psychology of Religion* 18: 169–194.

Schloss, J. P. and M. J. Murray. 2011. Evolutionary accounts of belief in supernatural punishment: A critical review. *Religion, Brain and Behavior* 1: 46–99.

Sears, D. O. 1986. College sophomores in the laboratory: Influences of a narrow data base on social psychology's view of human nature. *Journal of Personality and Social Psychology* 51: 515–530.

Shariff, A. F. A. 2011. Big gods were made for big groups. *Religion, Brain and Behavior* 1(1): 89–93.

Shariff, A. F. A. and A. Norenzayan. 2007. God is watching you. *Psychological Science* 18(9): 803–809.

Shariff, A. F. A. and A. Norenzayan. 2011. Mean gods make good people: Different views of God predict cheating behavior. *International Journal for the Psychology of Religion* 21(2): 85–96.

Shaver, J. H. 2015. The evolution of stratification in Fijian ritual participation. *Religion, Brain and Behavior* 5(2): 101–117.

Shaver, J. H. and R. Sosis. 2014. How does male ritual behavior vary across the lifespan? *Human Nature* 25(1): 136–160.

Sierksma, F. 1963. Sacred cairns in pastoral cultures. *History of Religions* 2(2): 227–241.

Slone, D. J. 2004. *Theological incorrectness: Why religious people believe what they shouldn't*. New York: Oxford University Press.

Smetana, J. G. 2006. Social-cognitive domain theory: Consistencies and variations in children's moral and social judgments. In J. Killen and E. Turiel, eds, *Handbook of moral development*. Mahwah, 119–153, NJ: Lawrence Erlbaum Associates.

Smith, E. A. 2000. Three styles in the evolutionary study of human behavior. In L. Cronk, W. Irons, and N. Chagnon, eds, *Human behavior and adaptation: An anthropological perspective*, 27–46. Hawthorne, NY: Aldine de Gruyter.

Smith, E. A. and M. Wishnie. 2000. Conservation and subsistence in small-scale societies. *Annual Review of Anthropology* 29: 493–524.

Snarey, J. 1996. The natural environment's impact upon religious ethics: A cross-cultural study. *Journal for the Scientific Study of Religion* 35(2): 85–96.

Soler, M. 2012. Costly signaling, ritual and cooperation: Evidence from Candomblé, an Afro-Brazilian religion. *Evolution and Human Behavior* 33: 346–356.

Sosis, R. 2005. Does religion promote trust? The role of signaling, reputation, and punishment. *Interdisciplinary Journal of Research on Religion* 1: 1–30.

Sosis, R. 2006. Religious behaviors, badges, and bans: Signaling theory and the evolution of religion. In P. McNamara, ed., *Where God and science meet: How brain and evolutionary studies alter our understanding of religion, vol. 1: Evolution, genes, and the religious brain*, 61–86. Westport, CT: Praeger Publishers.

Sosis, R. 2009. The adaptationist-byproduct debate on the evolution of religion: Five misunderstandings of the adaptationist program. *Journal of Cognition and Culture* 9: 315–332.

Sosis, R. 2011. Why sacred lands are not indivisible: The cognitive foundations of sacralizing land. *Journal of Terrorism Research* 2: 17–44.

Sosis, R. and E. R. Bressler. 2003. Cooperation and commune longevity: A test of the costly signaling theory of religion. *Cross-Cultural Research* 37: 211–239.

Sosis, R. and J. Bulbulia. 2011. The behavioral ecology of religion: The benefits and costs of one evolutionary approach. *Religion* 41(3): 341–362.

Sosis, R. and J. Kiper. 2014. Religion is more than belief: What evolutionary theories of religion tell us about religious commitment. In M. Bergmann and P. Kain, eds, *Challenges to religion and morality: Disagreements and evolution*, 256–276. New York: Oxford University Press.

Sosis, R., H. C. Kress, and J. S. Boster. 2007. Scars for war: Evaluating alternative signaling explanations for cross-cultural variance in ritual costs. *Evolution and Human Behavior* 28: 234–247.

Sosis, R. and B. J. Ruffle. 2003. Religious ritual and cooperation: Testing for a relationship on Israeli religious and secular kibbutzim. *Current Anthropology* 44: 713–722.

Sperber, D. 1997. Intuitive and reflective beliefs. *Mind and Language* 12(1): 67–83.

Sponsel, L. E. 2014. Spiritual ecology: Is it the ultimate solution for the environmental crisis? *Choice* 51(08): 1339–1348.

Strassmann, B. I. 1992. The function of menstrual taboos among the Dogon. *Human Nature* 3: 89–131.

Strassmann, B. I., N. T. Kurapati, B. F. Hug, E. E. Burke, B. W. Gillespie, T. M. Karafet, and M. F. Hammer. 2012. Religion as a means to assure paternity. *PNAS* 109: 9781–9785.

Swanson, G. E. 1960. *The birth of the gods: The origin of primitive beliefs*. Ann Arbor: University of Michigan Press.

Taylor, C. 2007. *A secular age*. Cambridge: Belknap Press of Harvard University Press.

Turiel, E. 1983. *The development of social knowledge: Morality and convention*. Cambridge: Cambridge University Press.

Wallace, A. F. C. 1966. *Religion: An anthropological view*. New York: McGraw-Hill.

Watanabe, H. 1972. The Ainu. In M. G. Bicchieri, ed., *Hunters and gatherers today: A socioeconomic study of eleven such cultures in the twentieth century*, 448–484. Prospect Heights, IL: Waveland Press.

Waters, F. 1963. *Book of the Hopi*. New York: Penguin Books.
Watts, F. and L. P. Turner. 2014. *Evolution, religion, and cognitive science: Critical and constructive essays*. New York: Oxford University Press.
Waytz, A., K. Gray, N. Epley, and D. M. Wegner. 2010. Causes and consequences of mind perception. *Trends in Cognitive Sciences* 14: 383–388.
White, L. 1967. The historical roots of our ecologic crisis. *Science* 155(3767): 1203–1207.
Whitehouse, H. 2004. *Modes of religiosity: A cognitive theory of religious transmission*. Walnut Creek, CA: AltaMira Press.
Xygalatas, D., P. Mitkidis, R. Fischer, P. Reddish, J. Skewes, A. W. Geertz, … J. Bulbulia. 2013. Extreme rituals promote prosociality. *Psychological Science* 24(8): 1602–1605.
Zuckerman, P. 2008. *Society without God: What the least religious nations can tell us about contentment*. New York: NYU Press.

9

Remembering Past Lives

Claire White, Robert Kelly, and Shaun Nichols

Introduction

In contemporary western contexts, and beyond, many adults have deep-seated convictions that they have lived before (e.g., Gallup 2005; Haraldsson 2006; Bender 2007).[1] If people really do have previous lives, that is astonishing; but even if they do not, the belief in past lives is still a striking phenomenon. What leads people to think that they existed before their birth?

The reality of past lives is often defended by appeal to external verifiable facts—a person remembers some particular distinctive previous-life event, which is then verified to have occurred. This idea is often championed in modern media (e.g., Wallis 1999; Winfrey 2013; Moye 2014). Parapsychology researchers have proposed a criterion to measure the validity of past-life memories based on similar assumptions (e.g., Stevenson 2001; Tucker 2005; Haraldsson 2006; Edelmann and Bernet 2007; Lyons 2013). The most famous of these researchers is Ian Stevenson. In an effort to provide scientific evidence for reincarnation, Stevenson investigated case studies of children who claim to have lived before. His methods included recording the child's recollection of people or events and then comparing their account to the accounts of others, which served as independent evidence of the accuracy of the event. For instance, Stevenson reports the case of Ravi Shankar Gupta, who recalled an experience of eating guavas before being taken to the riverside where he was murdered. This description was consistent with the murder of a boy named Munna Prasad, and so this was taken as evidence that the murdered boy was reincarnated as Gupta (Stevenson 1974, 96–97). In another case reported by Stevenson (1978), Shamlinie Prema reported remembering going to buy bread before school and falling into a flooded patty field; this cohered with the drowning death of Hemaseelie Guneratne and was again taken as evidence of reincarnation.[2]

Returning to the question of why people believe that they have had past lives, one possibility is that they, like the parapsychology researchers, also reason in terms of the external validation of their apparent episodic memories. It might be that when some people have an anomalous apparent memory (e.g., a dream or flash back) that contains very specific information that is clearly not from their own current life, this leads them to investigate whether this detailed information is accurate. For example, a person who seems to remember the inside of a WW1 plane, but has never learned about such planes, might seek out evidence to determine whether the apparent memory conforms to the actual layout of such planes. If external validation is found, this might provide the grounds for a conviction that the apparent memory was actually a real memory of a past life.

An alternative explanation for past-life beliefs is that the memory itself, independent of external validation, helps to generate or reinforce the belief that one existed at the time of encoding. In everyday life, a memory of past experiences can carry with it a sense of identity. My memory of experiences in Belfast from several years ago gives me the sense that *I* was in Belfast several years ago. The *kind* of memory that is involved here is *episodic memory*—memory of experiences. As Harlene Hayne and colleagues put it, episodic memory is "accompanied by conscious awareness that the event happened to 'me' or will happen to 'me' that does not accompany retrieval of other kinds of memories" (2011, 344). Insofar as episodic memory carries this sense that the event happened to "me," it also (of course) gives a basis for thinking that *I existed* at that time. We find this idea expressed already in the eighteenth century by Thomas Reid. He writes, "Our own personal identity and continued existence, as far back as we remember anything distinctly … we know immediately, and not by reasoning. It seems, indeed, to be part of the testimony of memory. Everything we remember has such a relation to ourselves as to imply necessarily our existence at the time remembered" (Reid 1785/1969, 586). So, in the case of the belief in past lives, one might have an apparent episodic memory of experiences dated before one's current life, and this memory, like episodic memories generally, might carry with it the sense of personal identity.

We have sketched two ways that episodic memory might contribute to the conviction that one has lived before. On the *external validation* approach, an apparent memory does not directly contribute to the conviction that one existed in a past life. Rather, the apparent memory can provide an impetus for determining whether there is independent reason to think that information contained in the memory is best explained by the hypothesis that one had a past life. The other way that (an apparent) memory might contribute to a belief in a past life is much more direct. On this view, episodic memory contributes to a past-life belief like *I was in Edinburgh in 1842* in much the way episodic memory contributes to the current-life belief *I was in Belfast in 2007*. In both cases, the (apparent) episodic memory of a past experience brings with it a sense of personal identity with the person who had that past experience. Of course, an episodic memory might carry both a sense of personal identity and also a motivation to seek external validation (see, e.g., descriptions of past-life memories in Table 9.1). So, the two roles of memory can work together, but they also provide independent tracks to the conviction of a past life.

Table 9.1 Most convincing type of memory and corresponding descriptions

Memory type	Memory description
Episodic	I remember dying in a ship. Something comes crashing by my left side, breaks through the wall and next I am sitting on the floor, I know something is wrong with my legs, I can't move and I start to feel cold . . . I bleed to death.
Episodic	I remember being in a foxhole clutching my Karabiner 98k rifle with artillery shells going off all around me. I remember feeling terrified and trying to mentally prepare myself for the fact I might not make it out of this.
Episodic	I was walking toward a castle that I could see in the distance. My clothes were odd, with vivid colors and I especially remember the shoes with curled tips. Suddenly a man on a horse drove quickly past me, and forced me out of the way.
Semantic	Knowledge of a time and place that I haven't been to this lifetime . . . I know way too much about the ocean and ships and pirates and beautiful gowns and love and the food of Europe.
Semantic	I identified a secret hidden place under the altar in the church. I wrote it down in the guest book because I had been there in a past life. They located the hidden room three years later.
Semantic	I was able to verify some of the facts (things I did not have previous knowledge about).

Let us turn now to the representation of self in past-life beliefs. The self gets characterized in radically different ways in different fields and traditions, but one useful distinction is between *trait-based* conceptions of the self and *non-trait based* conceptions of the self. In social psychology, the concept of self is taken to involve a collection of psychological traits. For instance, Jason Mitchell writes:

> The self-concept refers to a person's understanding of what she "is like" as a person, that is, what personality characteristics she manifests, what idiosyncratic abilities and proclivities define her as an individual, and to what extent she regards herself positively (i.e. has high or low self-esteem). (2009, 247)

People certainly do have trait-based conceptions of themselves (e.g., see Conway 2005). For instance, people are good at identifying which traits they have; this even holds for severely brain-damaged patients who have suffered the destruction of episodic memory (Klein et al. 2004).

The second conception of self, the *non*-trait conception holds that the self is not constituted by distinguishing properties like personality or convictions or aspirations. We see this idea in a famous pronouncement from Reid: "I am not thought, I am not action, I am not feeling; I am something that thinks, acts, and suffers" (1785/1969, 341). What exactly the self *is* on the non-trait conception is much less clear. The self might be represented in ways similar to how *essences* are thought to be represented in ordinary cognition. People associate raccoons with a set of features, but those features do not *define* raccoonhood; rather, people seem to think of a raccoon in terms of an "essence placeholder" which is associated with the familiar observable traits of raccoons (Medin and Ortony 1989). Similarly, the self might be represented in terms of an essence placeholder rather than a set of traits (Nichols forthcoming).

As we have seen, episodic memory can generate a sense of personal identity with a past individual. But what is the notion of self implicated in this episodic sense of identity? Is it the trait-conception of self? A little mnemonic exercise indicates that it is not. Recall some emotionally important event from your distant past: your first kiss, your high school graduation, your first car wreck. If you are like most adults, the traits you have now are very different from the traits of that kisser, graduate, and bad driver. But, does this make it seem like that kisser or bad driver was not *you*? During the recollection itself, it seems like that distant bad driver was exactly *me*, even though I know that that individual had very different traits than I currently have. This suggests that episodic memory delivers a conception of the self that is not essentially tied to traits (see also Nichols 2014).

The aim of this chapter is to address the role of memory in past-life convictions. Although it is commonly accepted in the modern media—and popular western culture more generally—that people believe they have lived before because the memory contains detailed verifiable facts, little is known about how people actually reason about the veracity of their previous existence. To our knowledge, the current project is the most extensive research that probes the role of memory in past life convictions. More specifically, we explore two questions. First, to what extent does memory lead to past-life belief, and is this conviction based on external validation or via the episodic sense of personal identity contained in the memory? Second, what is the conception of self on which one's current self is taken to be the same as the past-life self?

In what follows, we propose that memory plays an important role in convincing people that they have lived before. Fundamentally, this memory is episodic insofar as it represents the event as happened to the encoder. This memory contrasts with *semantic* memory—memory of facts. I might have a semantic memory of the date of my birth even though I have no recollection of the experience of being born. In contrast to semantic memory, episodic memory records events as experienced by the self at a particular time/place (e.g., the memory of celebrating my sixteenth birthday at my mother's house). Further, we contend that it is the episodic sense of personal identity (rather than external validation) that typically contributes to the belief in a past life. Finally, we argue that the conception of self implicated in past-life belief is the non-trait conception. Of course, this fits naturally with the idea that the belief in a past life issues partly from the episodic sense of identity, since it is the non-trait conception of self that is implicated in the episodic sense of identity. Indeed, Reid himself ties together the non-trait conception of self and the role of memory in the belief that the self has persisted:

> My thoughts, and actions, and feelings change every moment—they have no continued, but a successive existence; but that self or I to which they belong is permanent, and has the same relation to all the succeeding thoughts, actions, and feelings, which I call mine ... But perhaps it may be said this may all be fancy without reality. How do you know—what evidence have you—that there is such a permanent self which has a claim to all the thoughts, actions, and feelings which you call yours? ... To this I answer, that the proper evidence I have of all this is remembrance, I remember that twenty years ago I conversed with such a person; I remember several things that passed in that conversation: my memory testifies, not only that this was done, but that it was done by me who now remembers it. If it was done by me, I must have existed at that time, and continued to exist from that time to the present. (1785/1969, 318)

Our proposal in this chapter can be construed as a psychological extension of Reid's point in the context of past-life belief. People assume that they are the same person as someone in earlier times despite large variations in the traits between those individuals, and they think this partly because of the testimony of their episodic memory. If our account is correct, then it makes the belief in a past life less bizarre, at least psychologically. For we are suggesting that the episodic sense of personal identity that led to your conviction that you existed at the time of your sixteenth birthday celebrations some decades ago is also involved in other people's convictions that they existed 200 years ago. This proposal is also situated in recent cognitive approaches to the study of religion and religious experiences, which suggests that extraordinary convictions are often underpinned by the ordinary processes of social cognition (e.g., see Lawson and McCauley 1990; Barrett 2000; Boyer 2001; Barrett 2007; for reincarnation, see White forthcoming a, forthcoming b).

Although there has been much speculation about what leads people to the conviction that they have lived before, to date, there is a dearth of research that investigates this by asking the people who actually hold these beliefs. To redress this, we recruited western adults who believe they have had at least one past life and conducted an extensive survey concerning the role of memory in their conviction.[3] Our sample is largely white, middle-class adults who can be characterized as contemporary western new age spiritual seekers. That is to say, our participants do not identify exclusively with a mainstream religion as is traditionally defined ("nones"); rather, they incorporate aspects of historical religious traditions—such as Hindu concepts of reincarnation—and consider themselves open to individual spiritual experiences (i.e., "spiritual seekers"). We restricted our sample to residents of contemporary western countries to reduce the likelihood that participants would respond based on established dogma concerning the self, memory, and reincarnation. In particular, regulations about reincarnation (e.g., who can be reborn as whom), and corresponding cultural practices, such as identification procedures (i.e., how to identify someone who has been reborn) are notably absent in such contexts. The genealogy of contemporary American expression of past-life convictions, which characterizes the majority of our sample, occurs within a framework of new age spiritual exploration and individualism, rather than any organized religion that affirms reincarnation (e.g., see Schmidt 2000, 2005; Fuller 2001; Albanese 2006; Bender 2007).

Our view about the relationship between memory and identity leads to four predictions concerning the mnemonic foundations of past-life convictions and their respective roles in convincing people that they have lived before. First, we predict that most participants who hold the conviction that they have lived before will claim that a memory, rather than another personal feature (e.g., a physical mark or behavior), provides them with the most compelling evidence of their past-life existence. Second, we predict that most of our participants will characterize the single memory that most convinces them that they have lived before as an episodic rather than semantic memory. Third, we predict that more participants will claim that the feeling of exclusive ownership over the episodic memory characterizes their most convincing past-life memory, rather than the uniqueness of the information about their past life. Fourth, and finally, we predict that more participants will claim that the feeling of exclusive ownership over a single memory makes it their most convincing memory, rather than information it conveys about their unique personality or detailed information contained in it.

We asked participants three types of questions in the survey. First, to understand more about the role of memory and the nature of the self in past lives, we asked them about what convinced them that they had lived before. Second, to establish the degree of overlap between personality traits across lives, and memory ownership in past-life memories, we asked them which type of memory most convinced them of a past-life existence (i.e., episodic or semantic) and which properties characterized that memory. Third, to establish which property contributed most to their past-life conviction, we also asked participants to select which property characterized the memory that convinced them most of their past-life existence (e.g., information about the self, a sense of exclusive ownership of the memory). In what follows, we summarize these four key predictions:

The role of memory in past-life convictions:

Hypothesis 1: More participants will claim that memory, rather than other personal features (e.g., physical mark, personality trait), provides them with the most convincing evidence that they have had a past life.

Hypothesis 2: More participants will characterize the memory that most convinces them of their past life as an episodic memory—a memory of a personal experience—rather than a semantic memory concerning factual information.

Why memory provides evidence of a continuous self across lives:

Hypothesis 3: More participants will characterize their most convincing past-life memory as associated with the feeling of personal exclusive ownership rather than as providing them with distinctive information about their previous life.

Hypothesis 4: When forced to reason about what aspect of their memory convinces them of a past life, more participants will claim that it is the feeling of personal exclusive ownership that most convinces them of their past existence, rather than distinguishing information (e.g., about personality traits) contained in the memory.

Study 1

The survey was designed using the Qualtrics© survey builder and accessed online via an electronic link to a survey titled "Past Life Survey." The link to the survey was posted on past life and associated forums and websites, most often accessed by people who are interested in spiritual exploration and who do not identify with a world religious tradition.[4] The study was described to participants as "an investigation into why people think they have lived before." Participants had to be at least 18 years old, reside in a western country, and think that they had lived before their current biological lifetime. Demographic information was collected, but all surveys were anonymous. Ethical approval was obtained from The Committee for the Protection of Human Subjects at California State University, Northridge.

Method

Participants

A total of 195 adults volunteered for the study. However, thirty-five responses were excluded because they did not meet the inclusion criteria (e.g., lived in a non-western country). Therefore, the data of 160 participants were included in the study. The majority of participants were Caucasian (76.9 percent; other ethnicities: 23.1 percent) females (66.9 percent; male: 33.1 percent); many were in full-time employment (38.1 percent; employed: 25 percent; unemployed: 10.6 percent; students: 11.9 percent; retired: 8.1 percent; full-time parent: 4.4 percent; other: 1.9 percent). The mean age of participants was 42 years (SD = 14.172; range = 18–76). Most participants resided in the United States (71.3 percent; UK: 11.3 percent; other European countries: 6.3 percent; other western origins: 11.1 percent). More participants self-identified as *none* (i.e., not belonging to a major world religion: 37.5 percent) than any other religious affiliation (Buddhism: 7.5 percent; Christianity: 26.3 percent; Hinduism: 1.9 percent; Judaism: 1.9 percent; New Age/Spiritualist: 22.5 percent; other: 2.5 percent). When presented with a modified version of Thalborne's afterlife belief classification[5] (Thalbourne 1996), most participants (80 percent) described their belief about what happens after death as *reincarnation* (i.e., "after death, the person continues to exist and is reborn—perhaps many times"), and some (15 percent) identified as *unsure* as to what happens to the person after death (i.e., "after death, I am unsure what happens to the person").

A few participants (3.1 percent) characterized their beliefs as *eclectic* (i.e., "the person continues to exist and the person ceases to exist") or identified as believing in *resurrection* (1.9 percent, i.e., "the person continues to exist and their body may be resurrected"). These demographics, though skewed in terms of ethnicity, sex, religious affiliation, and afterlife belief, are typical of "nones" and "new age" spiritual seekers, which characterize the majority of our study population.[6]

Materials and procedure

The study was divided into two parts. In the first section, we asked participants whether memory (or another personal source) convinced them that they had lived before, and in the second section, we asked participants to reason about a single memory that they found most convincing as evidence of this belief.

Role of memory[7]

Past-life evidence

The basic assumption underpinning this research is that episodic memory, rather than other personal features (e.g., behavior, personality trait, physical mark), provides people with the most convincing source of personal evidence that they had lived before. Yet, this has never been established in research. It could be the case that, for example, participants regard a distinctive physical mark (e.g., a mole) as the most compelling personal evidence. Thus, the first question of interest is whether our participants do, in fact, think that a memory they regard as derived from a past life most convinces them that they lived before, rather than some other personal feature. We asked participants to select which one of the five personal features most convinced them that they had lived before: *memory, behavior, personality trait, physical mark*, and, to capture any recurrent feature that we had not accounted for in the forced choice options, *other*. Examples of the features other than *memory* were included because ethnographic records suggest that people also use these features to establish the past-life identity of other people in reincarnationist traditions (White 2015, forthcoming a, forthcoming b).

Judgments about the most convincing past-life memory

Participants were instructed to think about the memory that most convinced them that they had lived before, to describe the memory in 100 words or less, and to answer the remaining questions based on this single memory.

Memory type

Participants selected the statement that best described the memory from two possibilities. The first described the memory as *a personal experience that happened at a particular time and place* (an episodic memory), and the second was *a memory of a general fact* (a semantic memory).[8]

Properties present in most convincing memory

Participants selected all statements that they considered to be true about the single memory from a list of seven. The statements included self-referential properties: (a) *I feel as though I own this memory, and it belongs to no one else but me*, (b) *I can reexperience this memory*, and (c) *This memory makes me feel emotional (e.g., sad, happy) when I recall it*; content: (d) *This memory communicates the type of person I am*, (e) *Includes someone else who was, or is, important to me*, and (f) *This memory contains detailed information (e.g., exact time of day, details of location)*; and, included for breadth: (g) *Someone else told me that this was a past-life memory*, and (h) *None of these statements are true.*

Property that makes the single memory most convincing

Participants then selected, from the same list as above, one statement that they considered to be most convincing that they had a past life.

Results

Role of memory

As displayed in Table 9.2, and as predicted by hypothesis 1, more participants claimed that memory, rather than other personal features (e.g., behavior, personality trait, physical mark), provided them with the most convincing evidence that they had lived before. To determine whether the difference in proportion between memory and other features was greater than chance, we conducted a chi-square goodness-of-fit test. Distribution was not equal and the proportion of participants who selected memory (a majority of our sample, n = 104, 65 percent) was significantly different from a value of 20 percent (n = 32) that would be obtained by chance (i.e., equal distribution), χ^2 (4, n = 160) = 162.25, $p < .001$. This result is, to our knowledge, the first to confirm what previous researchers have assumed, that people who think they have lived before have memories of their existence at that time and find those memories the most convincing source of evidence over other personal features.

Table 9.2 Single feature that provides participants with the most compelling evidence of past-life existence

Trait	Number of participants	Percent of participants
Memory	104	65
Personality Trait	30	18.8
Other	20	12.5
Behavior	9	5.6
Physical Mark	4	2.5

a. Several participants (n = 32) selected *other* and then completed a 100-word free response describing the trait most convincing of having had a past life. A content analysis of these responses revealed that eight participants had described a memory and four specified that memory plus something else (e.g., behavior, all other traits, or a "Zen Master") most convinced them. The results of the content analysis were used in deriving the numbers in this table. However, only those participants who actually selected *memory* (n = 92) received further questions regarding their past-life memory, while all other participants were taken to the end of the survey. Thus, the total number of participants for the remaining data analyses was lower than that displayed here in Table 9.2 as it excluded those participants (n = 12) who only described a memory after selecting *other*.

Judgments about the most convincing past-life memory

As we are interested primarily in the role of memory in past-life convictions, participants who did not select *memory* (n = 68) as the feature that most convinced them that they had lived before were excluded from remaining analyses. Those who selected *memory* (n = 92; refer to note *a.* in Table 9.2) were included.[9]

Memory type

As predicted by hypothesis 2, more participants—and in fact, the vast majority of our sample (n = 86, 93 percent)—characterized their most convincing past-life memory as episodic (i.e., *a personal experience that happened at a particular time and place*) rather than semantic (i.e., *a memory of a general fact*), 7 percent (n = 6). (For examples of participants' descriptions of their memory, see Table 9.1.) To determine whether the distribution of participants who selected episodic memory was significantly higher than chance, we ran a chi-square goodness-of-fit test. The test indicated that the proportion of episodic past-life memories (n = 86, 93 percent) was significantly different from a value of 50 percent (n = 46) that would be obtained by chance, χ^2 (1, n = 92) = 69.57, $p < .001$. Again, this result is, to our knowledge, the first to confirm what previous researchers have assumed, that people who think they have lived before are relying most upon memories of personal events rather than memories of general facts about their past life.

Properties present in most convincing memory

As displayed by Figure 9.1, and as predicted by hypothesis 3, more participants felt that their memory was characterized by their exclusive ownership over it rather than information about their personality or detailed information about their past life. To determine whether this distribution was significantly greater than chance, we ran a chi-square goodness-of-fit test. There were two other properties besides exclusive ownership that more than half of participants ascribed to their memory (see Figure 9.1). However, since the main question of interest was the degree to which participants characterized their memory as conveying a sense of exclusive ownership as compared to information about their current or past self, a chi-square goodness-of-fit test was ran only for exclusive ownership. The test indicated that the proportion of participants who characterized the memory by their exclusive ownership over it (n = 85, 92.4 percent) was significantly different from a value of 50 percent (n = 46) that would be obtained by chance, χ^2 (1, n = 92) = 66.13,

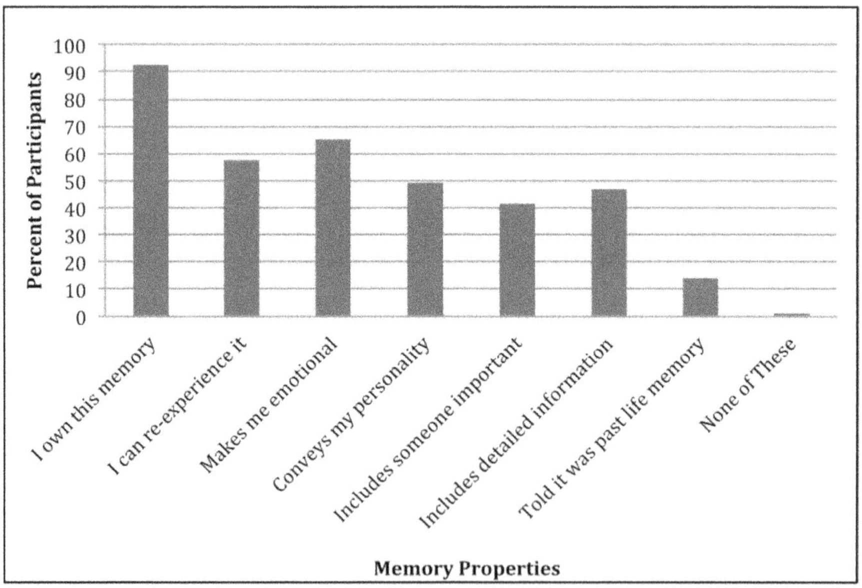

Figure 9.1 Memory properties present in most convincing memory.

p < .001. Overall, most participants characterized the past-life memory by the property and relations of the memories rather than the memory content. A majority claimed that they were able to reexperience the memory (n = 53, 57.6 percent), and that it made them feel emotional (e.g., happy, sad) upon retrieval (n = 60, 65.2 percent). In contrast, just less than half of the participants claimed that the memory conveyed something about their personality (n = 45, 48.9 percent), and that it contained detailed information about their past life (n = 43, 46.7 percent).

Property that makes the single memory most convincing

As Figure 9.2 demonstrates, and as predicted by hypothesis 4, when deciding which memory property led them to select it as the most convincing evidence of their past life, more participants selected the feeling of exclusive ownership over the memory as compared to any other property. We ran a chi-square goodness-of-fit test to determine whether this distribution was significantly greater than chance. The result indicated that the proportion of participants who said feeling as though they owned the memory most convinced them

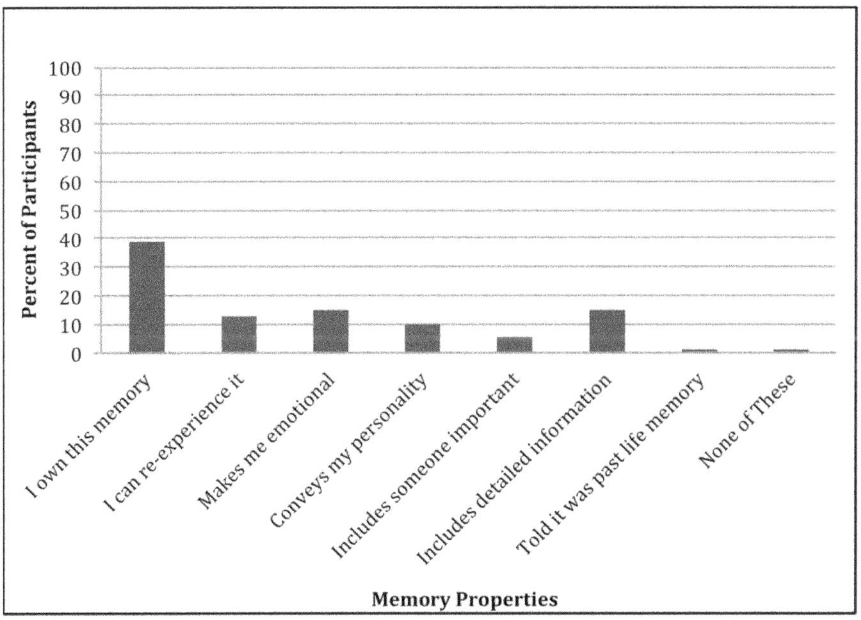

Figure 9.2 Memory property that makes the single memory most convincing.

of their past life ($n = 36$, 39.1 percent) was significantly different from the value of 12.5 percent ($n = 11.5$) that would be obtained by chance (i.e., equal distribution among the possible properties), $\chi^2 (1, n = 92) = 59.65, p < .001$. Of further interest was the finding that a similar number of participants found the ability to reexperience it ($n = 12$, 13 percent) and the emotions evoked upon recall ($n = 14$, 15.2 percent) as the most compelling aspect. Thus, participants find both the ability to reexperience the memory and the relationship toward the memory—even if this relationship is an association based on a single intense emotion (e.g., past-life regression therapy)—as evidence of their past life. Finally, more participants found the most compelling aspect of the memory that it contained detailed information about their past life ($n = 14$, 15.2 percent) than that it communicated something about their personality ($n = 9$, 9.8 percent), although the percentages for both are relatively low. We suspect that a priming effect has increased the number of participants who claimed that the detailed information contained in the memory made it most convincing. Namely, the question asks participants to select their most convincing past-life memory and thus explicitly to reason about what counts as evidence. Despite this bias, however, comparatively few participants (less than 16 percent) selected either of these choices (i.e., personality or detailed information). Thus, the result suggests that what constitutes reliable evidence in the metaphysical world and information about a person contained in the memory plays a relatively minor role in convincing people of the veracity of their past-life memory as compared to other aspects.

Discussion

The results of Study 1 supported all four hypotheses. First, more participants claimed that memory provided them with the most compelling evidence when compared to other sources (e.g., physical mark, personality trait). Second, more participants characterized their most convincing memory as episodic (i.e., a personal experience) rather than semantic (i.e., a fact). Third, more participants characterized this episodic memory as having the feeling of personal exclusive ownership over it rather than providing them with information about their personality or containing detailed information about their previous life. Fourth, and finally, when forced to reason about their most convincing past-life memory, more participants claimed that the feeling of personal exclusive ownership most convinced them of their previous existence rather than that the memory provided them with information about their personality or contained detailed information about their previous life. The magnitude of participants who selected the predicted choices is noteworthy. In most cases, this was a majority of participants; for all questions, the proportion of participants who selected the predicted option was significantly greater than proportions that would have been obtained by chance.

The results of Study 1 support our proposal that belief in a past life is supported by the episodic sense of personal identity. Overall, participants tended to characterize past-life memories by the experiential features of episodic memory (i.e., exclusive ownership, the ability to reexperience the memory and an emotional reaction upon retrieval of the memory) rather than the informational content of the memory. Further, participants tended to find these self-referential aspects most convincing of their past-life existence rather than the informational content of the memory. More participants claimed that the sense of exclusive ownership over a single memory both characterized their past-life memories and provided them with the sense that they existed at the time of encoding than did any other aspect of the memory. Although around half of the participants claimed that their past-life memory contained detailed information, only a few (n = 14, 15.2 percent) found it the most convincing aspect.

Our first study supported all of our hypotheses. For our second study, we wanted to explore two further issues concerning past-life belief. While our first study indicated that episodic memory plays a key role in past-life belief, the study did not explore the *perspective* of the memory and, therefore, the sense of identity implicated in it. Experiential memories can be recalled from the field (first person) or observer (third person) perspective (e.g., Nigro and Neisser 1983). When recalling an event from the field perspective, you see the scene from your original point of view; when recalling an event from the observer perspective, you see the scene as an observer might, for example, as from above. Some kinds of memories, for example, of tasting something delicious, are typically from the field perspective; others, for example, of swimming laps in a pool, are often recalled from the observer perspective as from above the pool (Nigro and Neisser 1983, 471). Obviously, our actual experiences are from the first person perspective, so it is widely assumed that it is through reconstructive processes (see, e.g., Loftus and Palmer 1974) that people come to have memories from an observer perspective (e.g., McIsaac and Eich 2004, 248). Although memories can be recalled from either the field or observer perspective, there are some qualities typically associated with these two perspectives. For instance, people recalling from the observer perspective are more likely to mention personal appearance, and those recalling from field perspective are more likely to use first-person pronouns (McIsaac and Eich 2002, 147–148). Most importantly for our purposes, when people remember scenes from the observer perspective, they are more likely to say that it does not *feel* like the event happened to them (e.g., Sierra and David 2011; cf. Klein and Nichols 2012).[10] The episodic sense of identity with the original experiencer seems to be significantly diminished in the observer perspective. Thus to the extent that the sense of personal identity with a past-life person comes from the episodic sense of identity, we should expect that the memory should be recalled from the field perspective rather than the observer perspective.

A second question that is not addressed by Study 1 concerns the conception of the self in past-life beliefs. Do people who believe that they have had a past life think that trait-similarity is essential to make them identical to the past person? Insofar as the belief in a past life depends on episodic memory, we might expect that the belief in a past-life does not depend on thinking that the previous individual is uniquely similar in traits.

To investigate these questions, and to further refine the extent to which self-trait knowledge provides a critical sense of identity across past lives, we conducted another study. For Study 2, we once again investigated the extent to which episodic memory plays an important role in past-life belief. This included an assessment of whether convincing past-life memories are from the field or the observer perspective. In addition, participants indicated the extent to which their current self had the same characteristics as the past-life self. We anticipated that our participants would think that there would be significant trait similarity between their past self and current self, but if trait similarity is what primarily drives participants' decisions that they are one and the same across lives, then we would expect the overlap to be uniquely strong, such that the traits themselves are sufficient to explain why they are the same person as the person in the past life. If that is the case, then participants should think that they are significantly more trait-similar to their past self than to any present person. To investigate this possibility, we introduced a comparison group to the past-life self—the closest living person to the participants. The main questions of interest, therefore, were whether the degree of similarity between the two (past and present self) would be substantial and significantly higher than the degree of similarity between the control group (present self and closest living person to them).

Predictions

Our general proposal is that past-life beliefs are typically generated and supported by the sense of personal identity issued by episodic memory. This leads to two hypotheses for the current study:

Hypothesis 1: More participants will characterize their most convincing past-life memory as recalled from the field (first person) perspective than the observer perspective.

Hypothesis 2: There will be no significant difference between the overlap in perceived trait similarity between (a) the present self and past-life self and (b) the present self and closest living person.

Study 2

Like Study 1, the survey was designed using the Qualtrics© survey builder and accessed via an electronic link titled "Past Life Survey 2." Again, the link to the anonymous survey was posted on past life and associated forums and websites and described as "an investigation into why people think they have lived before." The exclusion criteria were similar to Study 1 (i.e., participants had to be at least 18 years old, reside in a western country, and think that they had lived before their current biological lifetime), with the additional criteria that (1) participants had to regard a personal memory as the most convincing evidence that they had lived before (as compared to other personal features), and (2) they could not participate if they had participated in Study 1.

Method

Participants

A total of 104 adults participated in the study (eight were excluded because they did not meet the inclusion criteria). The demographics were similar to participants in Study 1. The majority of participants were Caucasian (81.7 percent; other ethnicities: 18.3 percent) females (71.2 percent; male: 28.8 percent) in full-time employment (51 percent; retired: 28.8 percent; other: 22.2 percent). The mean age of participants was 53.97 years (SD = 10.594; range = 27 to 72). Participants resided in the US (87.5 percent) or the UK (12.5 percent). More participants self-identified as New Age/Spiritualist than any other option (44.2 percent; none: 23.1 percent; Christianity: 23.1 percent; other: 9.6). Most participants described their belief about what happens after death as *reincarnation* (85.6 percent; *unsure*: 10.6 percent; *resurrection*: 3.8 percent).

Materials and procedure

The survey concerned whether and how similar the past and present self were, as evidenced by memory. Participants were asked to reason about a single memory that they found most convincing as evidence that they had lived before and to answer all questions based upon this single memory.

Memory type

To determine whether a new set of participants (compared to Study 1) would also tend to characterize their most convincing past-life memory as episodic, participants were asked to select the statement that best described the memory from two possibilities. The first described the memory as *a personal event that was experienced by me* (i.e., an episodic memory) and the second was described as a memory including *facts about the past that I could only know if I had experienced them* (i.e., a semantic memory).[11]

Field/observer

We provided a description of the distinction between recalling a memory from the field perspective as opposed to the observer perspective. To ensure that participants understood this distinction, we performed a two-part comprehension check for each perspective. Participants were given descriptions of two different memories and were asked to indicate whether they thought the memory was being recalled from the field perspective or the observer perspective. Only participants who passed both comprehension checks ($n = 55$, 69 percent) were included in the analyses for the question concerning memory recall perspective.[12] Following the comprehension checks, participants were asked to indicate whether their past-life memory was recalled from a field or observer perspective: "When you recall this memory, from which perspective do you recall it? If both, from which perspective is it most predominantly recalled? (Please explain briefly in the space provided.)"

Shared personal characteristics

Participants were instructed to think about all of their important characteristics in the present—their personality, temperament, major likes and dislikes, beliefs, values, ambitions, life goals, and ideals—and to answer the remaining questions based on these characteristics. First, to establish a baseline of how likely the participants were to share all of their important characteristics with other people in their current lifetime, participants were asked to list the closest living person to them. They then indicated the extent to which they shared all of their important characteristics with this person, based on a sliding scale from 0 (*we share none of these important characteristics*) to 100 (*we share all of these important characteristics*). Second, participants were asked to estimate the extent to which their current and past-life self shared these important characteristics, again on a sliding scale from 0 (*we share none of these important characteristics*) to 100 (*we share all of these important characteristics*).

Results

Memory type

Similar to findings in Study 1, most participants in Study 2 (n = 81, 80.2 percent) described their most convincing past-life memory as episodic (i.e., a personal event that they experienced), rather than semantic (i.e., a memory of facts about the past: n = 20, 19.8 percent). A chi-square goodness-of-fit test indicated that the proportion of episodic past-life memories (n = 81, 80.2 percent) was significantly different from a value of 50 percent that would be obtained by chance, χ^2 (1, n = 101) = 36.842, p < .001. As the focus of the study is on the role of episodic memory in past-life convictions, participants who did not select episodic memory (n = 20) were excluded from the main analyses, and only those who had described their memory as episodic (n = 81) were included.[13]

Field/observer

Sixty-nine percent of participants (n = 55) passed the comprehension questions for the field/observer distinction. Our analysis focused on these participants, and the data showed that a greater proportion of participants recalled their past-life memory from the field perspective rather than an observer perspective. A chi-square goodness-of-fit test indicated that the proportion of participants who recalled this memory from the field perspective (n = 38, 69.1 percent) was significantly greater than a value of 50 percent that would be obtained by chance, χ^2 (1, n = 55) = 8.018, p =.005.

Shared personal characteristics

As expected, many participants (n = 35, 43.2 percent) claimed that they shared many or all (i.e., 75–100 percent) important characteristics with their past-life self. However, many participants (n = 32, 39.5 percent) also claimed that they shared most of their important characteristics with the closest living person to them (see Table 9.3). A paired samples t-test indicated that there was no significant difference between the degree to which participants shared their important characteristics with their past-life self (M = 64.93, SD = 27.431) and with the closest living person to them (M = 63.49, SD = 27.645), $t(80) = .347$, $p = .730$. Thus, participants are as likely to think that another person, who is close to them, but not numerically identical to them, also shares these characteristics. Although the participants do share important traits with their past life self, absolute trait similarity is neither essential nor unique to the relationship between selves. This leaves open the possibility that other aspects of the relationship between selves are more important to the sense of self across lives in memory.

The results of Study 2 reinforce and extend the results of Study 1. First, across both studies participants regard memories of personal events (i.e., episodic memory) as the strongest personal source of evidence for this conviction when compared to other sources (e.g., personality trait). Further, in Study 2, participants tend to report remembering the past event from the field perspective, which, unlike the observer perspective, is strongly associated with the sense that the experience happened to the *self*. In addition, the findings of Study 2 further refine our understanding of the extent to which self-trait knowledge provides a critical sense of identity across lives. In Study 1, past-life memories were often characterized as containing information about

Table 9.3 Characteristics shared with past-life self and closest living person

Percent of characteristics shared	Past-life self (number of participants)	Past-life self (percentage of participants)	Closest living person (number of participants)	Closest living person (percentage of participants)
All (100)	11	13.6	9	11.1
Many (75–99)	24	29.6	23	28.4
Moderate (50–75)	37	33.3	29	35.8
Some (25–49)	8	9.9	12	14.8
Few (0–24)	11	13.6	8	9.9

the self, but the information about the self was rarely regarded as the most convincing aspect of the memory. The results of Study 2 indicate further that participants do not think that what makes them identical with the past-life person is similarity of traits; for participants did not tend to think that their current self shared more traits with their past self than with a current friend.

Conclusion

The main contribution of this research is to show how memory figures into the belief that one existed in a past life. Across two studies, we found that memory plays a powerful role in the belief in a past life. Moreover, we found support for the idea that the key aspect of most past-life memories is the *sense of personal identity* that episodic memories carry, rather than the detailed informational content of the memories. This sense of personal identity operates in everyday contexts to issue the conviction of identity across the biological life span. Thus, our view suggests that the belief in past lives depends partly on a mechanism that characteristically produces belief in the identity of self across time. Furthermore, this research indicates that although people tend to represent the self as similar in past lives, self-trait knowledge does not provide a critical sense of identity across lives. During normal biological life, episodic memory yields a sense of personal identity despite the knowledge of dramatic changes in traits between the current self and the past self. This holds for past-lives as well. Participants in our studies did not think that they shared more traits with their past self than they did with their closest living person, such as a friend.

Of course, there are significant differences between everyday beliefs in personal identity across a decade and the conviction of past lives across centuries. In western culture, only a minority of people believe that they have had specific past lives. Our results do not speak to the important differences between the belief in personal identity during biological life and across biological lives. The process of belief formation is notoriously complex and depends on a wide range of motivational, social, and cognitive processes that go beyond episodic recollections (see, e.g., Spanos et al. 1991; Bender 2007). We have not even attempted to give a complete etiology for past-life beliefs. Indeed, we have not even tried to give an etiology of the episodic memories involved in past-life beliefs. But we would like to stress that the fact that a person has an episodic memory that carries a sense of identity does not show that the memory reflects a real past life. The reconstructive processes involved in memory fixation can lead to episodic memories that are false in detail (e.g., whether a car crash involved broken glass) or entirely inaccurate even about one's presence. Getting people to imagine vividly having done something can lead them to remember actually doing that thing (Goff and Roediger 1998; Thomas and Loftus 2002; Rajagopal and Montgomery 2011). We do, however, think it is important to explore these issues in participants who actually have convictions about specific past lives. Of course, there may be other anomalous experiences (e.g., Déjà vu, hallucinations, unusual physical sensations or out of body experiences) derived from sources other than memory recall (e.g., visualization in meditation, information provided by a psychic) that play an important role in people's convictions that they have lived before. There are also likely to be differences in how people access memories attributed to a past life (e.g., deliberate versus spontaneous recall). Our results do, however, suggest that the sense of personal identity delivered by episodic memory plays a central role in the convictions of these individuals. Overall, this research contributes to cognitive approaches to the study of religion, much of which suggests that "extraordinary" convictions, including those associated with reincarnation, are critically underpinned by some of the same processes that govern mundane social cognition (see, e.g., Lawson and McCauley 1990; Barrett 2000; Boyer 2001; Barrett 2007; White forthcoming a, forthcoming b).

Acknowledgments

This research was funded by a Research, Scholarship, and Creative Activity Award at California State University, Northridge. We would like to thank David Alpizar for his assistance with the statistical analyses in this chapter.

Notes

1 Around 20 percent of western residents believe in reincarnation (Gallup and Proctor 1982; Ashford and Timms 1992; Walter 2001). Those who believe they have had a past-life are subsumed into the category of "new age spiritual seekers" (see Bender 2007), and research polls indicate that new age spiritual seekers characterize about 8 percent of the US population (Heelas 2006).

2 Stevenson also used birthmarks as evidence of reincarnation, devoting a major tome to the issue (1997).

3 Although we are not the first to ask people about their past lives, our survey is the most extensive, and it is agnostic about the possibility of reincarnation. Most research on past-life convictions has taken place in western parapsychology and clinical psychology, where the aim of research is to test the reliability of these past-life memories (e.g., Stevenson 2001; Haraldsson 2006; Edelmann and Bernet 2007; Lyons 2013). On the one hand, parapsychological case studies across the world often highlight past-life memories in which the details of memory are confirmed against the odds (e.g., no prior contact with people they recognize, technical details of a time in history or about a different country). On the other hand, in clinical frameworks the focus is also on the reliability of past-life memories, typically with the assumption that the memories are false reconstructions. In clinical research, the aim is often to uncover individual differences in the susceptibility to false memory distortion (e.g., Barker and Pasricha 1979; Spanos et al. 1991; Roediger and McDermott 1995; Haraldsson 1997; Stevenson 2001; Meyersburg et al. 2009; Pyun and Kim 2009).

4 E.g., *Spiritual Awakening and Self-Realization* (https://www.facebook.com/groups/259686714183097/), *Remember Who You Are* (https://www.facebook.com/groups/183665478346811/).

5 We modified Thalbourne's original questionnaire to make it clearer to the participants. Specifically, the words "body," "conscious personality" and "soul" were replaced with "the person." In addition, we added "resurrection" as an option and dropped "immortalist." We also labeled "agnostic" here as "unsure," for purposes of clarity since the latter seems to more accurately reflect the statement. We also removed the sixth option in the original questionnaire, "other believer," since it did not seem to be distinct from the other options (e.g., unsure) given the study focus on the role of the body.

6 Preliminary analyses revealed no main effect of gender or religious affiliation on the participants' responses and so they were not considered further in the studies.

7 Participants did not see the section descriptions, which are given here for clarification.

8 Parenthetical notes (i.e., episodic or semantic memory) were not included in the survey but are provided here for clarification.

9 Demographics for the participants who selected "memory" as the most convincing piece of evidence (n = 92) was comparable to the original sample (n = 160) and was as follows: Caucasian (76.1 percent; other ethnicities: 23.9 percent) females (75 percent; male: 25 percent) in full-time employment (38 percent; employed: 25 percent; unemployed: 10.9 percent; students: 10.9 percent; retired: 8.7 percent; full-time parent: 5.4 percent; other: 1.1 percent). The mean age of participants was 42.46 years (SD = 14.32, range = 18 to 76). Most participants resided in the United States (71.7 percent) or the UK (10.9 percent). Also, most self-identified as *nones* (43.5 percent; Buddhism: 2.2 percent; Christianity: 30.4 percent; Judaism: 1.1 percent;

New Age/Spiritualism: 20.7 percent; other: 2.2 percent) and described their belief about what happens to the body after death as *reincarnation* (81.5 percent; eclectic: 3.3 percent; resurrection: 2.2 percent; unsure: 13 percent).

10 Indeed, people with posttraumatic stress disorder often remember the traumatic event as from an observer perspective, and this is thought to reflect a coping strategy since event memories are less emotional when recollected from the observer perspective (e.g., McIsaac and Eich 2004).

11 Again, parenthetical notes (i.e., episodic or semantic memory) were not included in the survey but are provided here for clarification.

12 As our main interest was the difference between participants recalling their memory from the field perspective versus the observer perspective, those who chose "other" (n = 4), but still passed both comprehension checks, were excluded from this analysis, although the difference also reached significance when they were included.

13 Owing to an oversight in the survey, in study 2, participants were given less information about what constituted a semantic memory than Study 1. Specifically, unlike Study 1, where the semantic memory option was explicitly described as "not an event," Study 2 did not include this exclusion criterion. As a result, we noted that some participants had selected "semantic" but seemed to describe an episodic memory (of an event experienced by the self, without mentioning facts learned through the memory). A coder, blind to the study hypotheses, re-characterized those who had selected the "semantic" option as either semantic or episodic based on their memory description. Seventeen participants had selected "semantic" but had described an episodic memory. They were included in the remaining analyses of 81 participants. Three participants from the initial sample of 104 participants were excluded from the analyses because they selected "semantic" and gave obscure descriptions that could not be coded as either semantic or episodic.

Bibliography

Albanese, C. L. 2006. *A republic of mind and spirit: A cultural history of American metaphysical religion*. New Haven, CT: Yale University Press.

Ashford, S. and N. Timms. 1992. *What Europe thinks: A study of Western European values*. London: Dartmouth Publishing Group.

Barker, D. R. and S. K. Pasricha. 1979. Reincarnation cases in Fatehabad: A systematic survey in North India. *Journal of Asian and African Studies* 14(3–4): 231–241.

Barrett, J. L. 2000. Exploring the natural foundations of religion. *Trends in Cognitive Sciences* 4(1): 29–34.

Barrett, P. 2007. Structural equation modelling: Adjudging model fit. *Personality and Individual Differences* 42(5): 815–824.

Bender, C. 2007. American reincarnations: What the many lives of past lives tell us about contemporary spiritual practice. *Journal of the American Academy of Religion* 75(3): 589–614.

Boyer, P. 2001. *Religion explained*. New York: Basic Books.

Conway, M. A. 2005. Memory and the self. *Journal of Memory and Language* 53(4): 594–628.

Edelmann, J. and W. Bernet. 2007. Setting criteria for ideal reincarnation research. *Journal of Consciousness Studies* 14(12): 92–101.

Fuller, R. C. 2001. *Spiritual but not religious: Understanding unchurched America*. New York: Oxford University Press.

Gallup, Inc. 2005. *Three in four Americans believe in paranormal*. http://www.gallup.com/poll/16915/Three-Four-Americans-Believe-Paranormal.aspx#2

Gallup, G. and W. Proctor. 1982. *Adventures in immortality*. New York: McGraw-Hill.

Goff, L. and H. Roediger III. 1998. Imagination inflation for action events: Repeated imaginings lead to illusory recollections. *Memory & Cognition* 26: 20–33.

Haraldsson, E. 1997. A psychological comparison between ordinary children and those who claim previous-life memories. *Journal of Scientific Exploration* 11(3): 323–335.

Haraldsson, E. 2006. Popular psychology, belief in life after death and reincarnation in the Nordic countries, Western and Eastern Europe. *Nordic Psychology* 58(2): 171–180.

Hayne, H., J. Gross, S. McNamee, O. Fitzgibbon, and K. Tustin. 2011. Episodic memory and episodic foresight in 3-and 5-year-old children. *Cognitive Development* 26(4): 343–355.

Heelas, P. 2006. The infirmity debate: On the viability of new age spiritualities of life. *Journal of Contemporary Religion* 21(2): 223–240.

Klein, S. B., T. P. German, L. Cosmides, and R. Gabriel. 2004. A theory of autobiographical memory: Necessary components and disorders resulting from their loss. *Social Cognition* 22(5: Special issue): 460–490.

Klein, S. B. and S. Nichols. 2012. Memory and the sense of personal identity. *Mind* 121(483): 677–702.

Lawson, E. T. and R. McCauley. 1990. *Rethinking religion: Connecting culture and cognition*. Cambridge: Cambridge University Press.

Loftus, E. F. and J. C. Palmer. 1974. Reconstruction of automobile destruction: An example of the interaction between language and memory. *Journal of Verbal Learning and Verbal Behavior* 13(5): 585–589.

Lyons, S. 2013, December 10. The science of reincarnation: U.Va. psychiatrist Jim Tucker investigates children's claims of past lives. *University of Virginia Magazine*. http://uvamagazine.org/articles/the_science_of_reincarnation

McIsaac, H. K. and E. Eich. 2002. Vantage point in episodic memory. *Psychonomic Bulletin & Review* 9(1): 146–150.

McIsaac, H. K. and E. Eich. 2004. Vantage point in traumatic memory. *Psychological Science* 15(4): 248–253.

Medin, D. L. and A. Ortony. 1989. Psychological essentialism. In S. Vosniadou and A. Ortony, eds., *Similarity and analogical reasoning*, 179–195. Cambridge: Cambridge University Press.

Meyersburg, C. A., R. Bogdan, D. A. Gallo, and R. J. McNally. 2009. False memory propensity in people reporting recovered memories of past lives. *Journal of Abnormal Psychology* 118(2): 399–404.

Mitchell, J. P. 2009. Social psychology as a natural kind. *Trends in Cognitive Sciences* 13(6): 246–251.

Moye, D. 2014, February 22. *Reincarnated kids: 9 signs your child had a past life*. http://www.huffingtonpost.com/2014/02/21/reincarnated-kids_n_4832071.html

Nichols, S. 2014. The episodic sense of self. In J. D'Arms and D. Jacobson, eds., *Moral psychology and human agency*, 137–155. Oxford: Oxford University Press.

Nichols, S. forthcoming. The essence of agents. Synthese.

Nigro, G. and U. Neisser. 1983. Point of view in personal memories. *Cognitive Psychology* 15(4): 467–482.

Pyun, Y. D. and Y. J. Kim. 2009. Experimental production of past-life memories in hypnosis. *International Journal of Clinical and Experimental Hypnosis* 57(3): 269–278.

Rajagopal, P. and N. V. Montgomery. 2011. I imagine, I experience, I like: The false experience effect. *Journal of Consumer Research* 38(3): 578–594.

Reid, T. 1969. *Essays on the intellectual powers of man*. Cambridge, MA: MIT Press. (Original work published 1785.)

Roediger, H. L. and K. B. McDermott. 1995. Creating false memories: Remembering words not presented in lists. *Journal of Experimental Psychology: Learning, Memory, and Cognition* 21(4): 803–814.

Schmidt, L. E. 2000. *Hearing things: Religion, illusion, and the American enlightenment*. Cambridge, MA: Harvard University Press.

Schmidt, L. E. 2005. *Restless souls: The making of American spirituality*. Berkeley: University of California Press.

Sierra, M. and A. S. David. 2011. Depersonalization: A selective impairment of self-awareness. *Consciousness and Cognition* 20(1): 99–108.

Spanos, N. P., E. Menary, N. J. Gabora, S. C. DuBreuil, and B. Dewhirst. 1991. Secondary identity enactments during hypnotic past-life regression: A sociocognitive perspective. *Journal of Personality and Social Psychology* 61(2): 308–320.

Stevenson, I. 1974. *Twenty cases suggestive of reincarnation*. Charlottesville: University of Virginia Press.

Stevenson, I. 1978. *Cases of reincarnation type [Volume II]: 10 cases in Sri Lanka*. Charlottesville: University of Virginia Press.

Stevenson, I. 1997. *Reincarnation and biology [2 Volumes]: A contribution to the etiology of birthmarks and birth defects*. New York: Praeger.

Stevenson, I. 2001. *Children who remember previous lives: A question of reincarnation*. Jefferson, NC: McFarland.

Thalbourne, M. A. 1996. Belief in life after death: Psychological origins and influences. *Personality and Individual Differences* 21(6): 1043–1045.

Thomas, A. K. and E. F. Loftus. 2002. Creating bizarre false memories through imagination. *Memory & Cognition* 30(3): 423–431.

Tucker, J. B. 2005. *Life before life: A scientific investigation of children's memories of previous lives.* New York: Macmillan.

Wallis, D. 1999, September 26. *Conversations/Dr. Ian Stevenson: You may be reading this in some future past-life.* http://www.nytimes.com/1999/09/26/weekinreview/conversations-drian-stevenson-you-may-be-reading-this-in-some-future-past-life.html

Walter, T. 2001. Reincarnation, modernity and identity. *Sociology* 35(1): 21–38.

White, C. 2015. Establishing personal identity in reincarnation: Minds and bodies reconsidered. *The Journal of Cognition and Culture*, 15, 403–430.

White, C. forthcoming a. Cross-cultural similarities in reasoning about personal continuity in reincarnation: Evidence from South India. *Religion, Brain & Behavior.*

White, C. forthcoming b. The cognitive foundations of reincarnation. *Method and Theory in the Study of Religion.*

Winfrey, O. (Host). 2013. Oprah and Dr. Brian Weiss: Reincarnation, past lives, and miracles [Television series episode]. In J. Barancik and C. Koehl (Producers), *Super Soul Sunday.* Chicago, IL: Harpo Productions.

10

The Death We Fear Is Not Our Own: The Folk Psychology of Souls Revisited and Reframed

K. Mitch Hodge

My fear of my death

A long-held and largely unchallenged assumption has been that we, humans, have and hold afterlife beliefs because we fear our own death. Exactly what we fear about our death has varied across interested parties—whether it be the fear of possible (even probable) annihilation, the process of dying, or fear of eternal punishment—but this fear's centrality to both the genesis and acceptance of afterlife beliefs has simply seemed obvious. According to more recent theorists, if we did not believe in some form of immortality, literal or symbolic, the immense fear of our own inevitable demise would be debilitating. We would be so terrified of dying that we would fail to do what is necessary to live.

In fact, it was to overcome one's fear of death and to prepare for one's own death that the ancients believed was the *true* purpose of philosophy. Some have even conjectured that overcoming the fear of death is the genesis and driving force behind all religion. More modestly, as I originally stated, the connection between humans' fear of their own death and the development and acceptance of afterlife beliefs seems academic. We believe in an afterlife because we cannot accept, let alone face, our inevitable death. We live in denial of death.

In this chapter, I will consider new empirical evidence that suggests humans naturally and intuitively believe in an afterlife, and that the genesis of this belief is independent from the fear of one's own death and cultural indoctrination. Even more surprising is that the current evidence suggests that one's development and acceptance of afterlife beliefs has little to do with oneself at all. It is not our own death, nor the fear of it, that provides the impetus for afterlife beliefs. I will certainly not go as far to say that the fear of our own death plays *no* role in the acceptance of afterlife beliefs: I will simply be arguing that this fear does not play the central and crucial role in their development and acceptance as has for so long been assumed.

Instead, I will argue from the evidence that afterlife beliefs are not egocentric, but allocentric. Our belief in an afterlife is spontaneously generated and intuitively accepted when we think about deceased others. Moreover, I will argue that this is not *necessarily* a traumatic or anxiety-laden event as has been for so long assumed. On the contrary, it relies on mundane cognitive systems that we use every day in dealing with, and thinking about, our conspecifics.

From the consideration of this evidence, I will ask how philosophers can assist in understanding death in relation to afterlife beliefs. Can philosophy finally provide the long sought-after consolation that it has so long promised? And, if the fear of our own death is minimized as a motivation for the genesis and acceptance of afterlife beliefs, what role is left for philosophy?

The philosophy of the fear of death and afterlife beliefs

I will not spend a great deal of time recounting how philosophers, from ancient to present day, have associated the fear of death, annihilation, or eternal reward or punishment with the genesis and acceptance of belief in an afterlife, especially since excellent reviews on those topics exist (for instance, see Benatar 2004; Martin and Barresi 2006; Sorabji 2006). What is of particular interest for my purposes here is how philosophers have suggested that we face our inevitable death: either by realizing that our annihilation at death is nothing to fear or by accepting (in the West) one of three alternative beliefs about how we survive death.[1]

The Epicureans believed that death is nothing to us—quite literally. Once we are dead, we are annihilated. Our self, including our thoughts, our hopes, our loves, and *everything* that makes us who we are, is extinguished at death. Where the fear arises for us is that we try to imagine how we would *feel* in such an annihilated state. But this, according to the Epicureans, is irrational. Being annihilated at death, we will no longer think or feel about anything. The Epicurean Lucretius argued that the rational way to think about our annihilation at death, and thus overcome our fear, was to view it the same way as we do our nonexistence prior to our birth (or conception). In the same way we had no thoughts or feelings then, and thus were not able to fear our nonexistence, our nonexistence after our death will be the same. Death is nothing to us.[2]

If one finds the prospect of her future annihilation troubling, fear not since alternative views on how we might survive death have been offered. In his definitive works on immortality, Flew (1953, 1956, 1967/2006, 1998) outlined the three most popularly held and widely debated views in the West on how we might survive our own death: The Platonic-Cartesian Way (PCW), The Reconstitutionist Way (RW), and The Astral Body Way (ABW).[3]

PCW has proved to be of the greatest interest and promise to philosophers attempting to explain how it is possible to survive death. Although PCW has been offered in various "strengths," what all PCW positions have in common is a soul-body dualism—that is, we humans (at least) in our current living form in this life are the mixture of two distinct substances: a soul and a body. Throughout the long history of defense for the PCW view, the soul substance has always been given primacy over the body. This soul substance has been described, from weakest to strongest, as our life-force, as our essence, as our "image" of God, as our moral center, and as our mind. Of these, the latter has proved to be the most popular among modern and contemporary philosophers. Although hints of this view appear much earlier in the history of philosophy, this view is credited and strongly associated with Descartes (1641/1993). After allowing himself to be overcome with hyperbolic skepticism about the existence of everything, including himself, Descartes famously reasoned, "*Cogito ergo sum*." From this conclusion, Descartes went on to prove that not only did he indeed exist, but so did God, other people, and the entire physical world. Most importantly for our present purposes, Descartes's arguments went on to conclude that he was a mind that occupied a beast-machine (his body); that the mind and the body were different independent substances (immaterial and material, respectively); that the mind survives the death of the body, and continues existing as an immortal, disembodied, rational entity. Descartes's assumption through all of these arguments is that the mind *is* the elusive soul for which philosophers had been searching. While Descartes delivered himself from his self-induced solipsism, his arguments were patently egocentric. It was from *his* existence and immortality that he demonstrated the existence and immortality of others.

RW should be, one would think, the most popular in the West, particularly among Christian philosophers. Philosophically speaking, however, it has received little attention. RW represents our survival of death as the literal reconstitution of our earthy bodies sometime after death by divine fiat.[4] This way, as Flew rightly points out, is the official afterlife doctrine of the Christian Church, even though few philosophers (including Church Fathers) have argued for this representation.[5]

Finally, ABW represents survival of death as our continued existence as some apparitional embodied form—and here is it perfectly acceptable to let your mind drift to the famous dance scene between Demi Moore and Patrick Swayze (as his deceased character) in *Ghost*, or representations of "afterliving deceased" in *The Sixth Sense* since those hit the nail right on the head with this type of representation. ABW claims "that a person is a kind of shadow man, sufficiently human and corporeal to overcome the problem of identification with the familiar flesh and blood person and at the same time sufficiently ethereal and elusive to have no difficulty in escaping unnoticed from the ordinary earthly body which is destined to be burned or buried (Flew 1967/2006, 603)." Additionally, Flew noted that ABW was undoubtedly the most commonly appealed to representation of how we survived death in psychical research and literature and other media but has rarely been seriously entertained by philosophers. For my purposes in this chapter, I will focus on the PCW and ABW.

To what I wish to draw the readers' attention at this point is that PCW does a better job of representing how it might be possible for me to conceive of my own survival of death; whereas ABW focuses more on how we, in this life, represent deceased others. If it is the case that securing one's own immortality in the face of the terror of one's inevitable death is the main motivation and impetus for afterlife beliefs, then PCW, given its egocentric genesis, should best fit with how one represents life after death. However, if it is the case that something other than the fear of one's own death is responsible for one's belief in an afterlife, then we should expect more of an ABW style of representations for reasons that I will provide shortly. As things stand now in philosophy, to the extent that afterlife beliefs are held and defended by theistic philosophers, the egocentric PCW way is by far the most popular and the most developed, and this is, in part, because philosophers have continued to assume that afterlife beliefs themselves are egocentric representations generated to achieve one's own immortality.

The science of the fear of death and the genesis of afterlife beliefs

A subfield has developed in contemporary psychology centered on the fear of (one's own) death, called Terror Management Theory (TMT). TMT psychologists took up the mantle of cultural anthropologist Ernest Becker's theory (1973) that because we humans are self-aware, we know that we are going to die.

This creates within each one of us an innate fear of death—understood as our *ultimate annihilation*. This knowledge of our ultimate annihilation would have debilitated humans if we had not found a way to overcome or ameliorate the terror which such knowledge brings. In other words, if self-aware humans had not found a way by which to deny death, humans would have been overcome by such extreme anxiety of their impending death that they would have failed to engage in activities necessary to sustain their (however brief) lives, and the human species would have become extinct before it even really began. According to Becker, that means was culture: specifically religious culture that provided for either our literal or symbolic immortality. Thus, afterlife beliefs are culturally derived because of our collective, yet individual, terror of our inevitable, ultimate annihilation (death). According to the psychology of TMT, when we are reminded of our inevitable annihilation at death (whether consciously or unconsciously), we use our cultural worldviews which provide us a means of immortality to assuage overwhelming anxiety of that certain outcome (Greenberg et al. 1990; Solomon et al. 1997; Vallacher 1997; Pyszczynski et al. 1999; Goldenberg 2005; Vail III et al. 2010).

It is not merely TMT theorists who think that humans' default way of thinking about death is as annihilation. As we saw earlier, philosophical positions regarding the possibility of an afterlife are spawned to overcome our understanding of death as annihilation. Even scientists who challenge TMT's assigned supporting role of religion and afterlife beliefs to overcoming the fear of death also refer to humans' default way of viewing death is as annihilation (Astuti and Harris 2008; Richert and Smith 2010; Astuti 2011; Bek and Lock 2011; Pereira et al. 2012). Additionally, studies by Jesse Bering (Bering and Bjorklund 2004; Bering et al. 2005), whose findings I will explore more below, assert from an early age that humans understand death as being inevitable, irreversible, and a *complete* cessation of *all* function which he assumes means we understand death as annihilation. This assumed concept of death is supported by studies targeting children's developmental understanding of (the consequences of) *biological* death, such as those by Barrett and Behne (2005) and Speece and Sandor (1984). Most certainly, if humans' default understanding was that all of us face complete annihilation at death, we would have every reason to be terrified!

As we have already seen, even though TMT offers a cultural genesis of afterlife beliefs, it is still based on an innate, egocentric fear of one's own annihilation. But, studies that attempt to provide a psychological/cognitive genesis for afterlife beliefs rely on egocentric theories as well, of which I will briefly discuss three, but of which I have also provided a much broader analysis in Hodge (2011b): they are Paul Bloom's Intuitive Cartesian Substance Dualism (2004, 2007), Jesse Bering's Simulation Constraint (2002, 2006), and Shaun Nichols' Imaginative Obstacle (2007).[6] These latter two rely heavily on one's ability to imagine/represent/simulate *oneself* as nonexistent and thus are clearly egocentric. Bloom's Cartesian substance dualism theory (2004) contains both allocentric and egocentric elements, but his language is exclusively egocentric when discussing children's afterlife beliefs. Bloom claims that infants (Kuhlmeier et al. 2004a, 2004b) intuitively and naturally understand themselves and others as immaterial mental beings who occupy physical bodies. In other words, they naturally, intuitively, and even unconsciously believe that the mind and body are independent of each other. But, when his attention turns to afterlife beliefs, Bloom claims that even though children can easily imagine the destruction of *their* body, "even young children should believe that the soul survives the destruction of the body (Bloom 2004, 207)." Bloom asserts that afterlife beliefs are the natural consequence of our intuitive Cartesian substance dualism because children believe that they can survive the destruction of *their own* bodies as immaterial minds. Bering claims that young children believe in an afterlife because, while they can imagine what is like for themselves not to eat, be thirsty or hear, they find it cognitively difficult to simulate what it would be like for themselves not to think, love, or desire. Bering too believes this is ultimately because children are naturally and intuitively mind-body dualists when it comes to themselves and thus can only simulate themselves as shed of bodily constraints but not mental ones. Finally, Nichols argues that we believe in an afterlife because we encounter a cognitive obstacle when we attempt to imagine our own future nonexistence. My mind encounters this obstacle when I imaginatively attempt to represent the statement, "It is the future and I do not exist." My imaginative abilities are limited in such a way as to prevent me from entertaining such egocentric statements—even though I could easily imagine it is the future and *you* are not here. Nichols, like Bloom and Bering, ultimately hangs his hat on some sort of egocentric intuitive dualism with regard to the genesis and acceptance of afterlife beliefs.

In this section, we have seen that in the science of afterlife beliefs, as it was with philosophy, the fear of our own death and other egocentric cognitive processes are proposed to play a central role in the genesis and acceptance of afterlife beliefs. In the next section, I will argue that the empirical evidence gathered from experiments investigating afterlife beliefs (especially in young children) challenges the centrality of both.

The folk psychology of souls[7]

Numerous experiments investigating both the intuitiveness and structure of afterlife beliefs have been conducted which all have a similar experimental design, whether conducted with adults (Bering 2002; Bek and Lock 2011; Huang et al. 2013), children (Bering and Bjorklund 2004; Bering et al. 2005; Harris and Giménez 2005), or both (Astuti and Harris 2008).[8] In each experiment, the participant was presented a narrative regarding a living subject *S* who meets a sudden death at the end of the story. Throughout the narrative, numerous physical and mental states were attributed to *S*: physiological (e.g., eating), psychobiological (e.g., thirsty), perceptual (e.g., smelling), epistemic (e.g., knowing), desirous (e.g., wanting), and emotional (e.g., loving). After hearing/reading the narrative, the participant was then asked a series of questions to determine whether she thought that *S* was still capable of performing those activities now that *S* was dead. The general format for those questions was, "Now that *S* is dead, can *S* still do activity *A*?" where *A* was a specific example from one of the aforementioned categories.

All of the experiments found similar results: Participants (across ages, religious backgrounds, educational levels and cultures) were significantly more likely to attribute epistemic, desirous, and emotional abilities to *S* after his death than they were to attribute physiological, psychobiological, or perceptual abilities.[9] For analysis, the former three abilities were lumped together as mental states, and the latter three were lumped together as physical states based on Theory of Mind psychology.[10] As a result of this division, most of the researchers claimed that the results supported some form of mind-body dualism which they aligned to varying degree with Bloom's (egocentric) view of intuitive Cartesian substance dualism. From this, the researchers claimed that both children and adults intuitively saw people in the afterlife as disembodied minds. In other words, their claim was that humans intuitively viewed deceased others in similar fashion of PCW as laid out by philosophers.[11]

Another interesting topic discussed in many of these experiments is the difference between the religious conception of death and the biological conception of death (Bering 2002; Bering et al. 2005; Harris and Giménez 2005; Astuti and Harris 2008; Bek and Lock 2011). Each of the researchers believed that there was inherent conflict with which our minds struggled when thinking about or imagining deceased individuals, particularly ourselves. On the one hand, they had suffered the inevitable and irreversible fate which brought about the cessation of all (biological) functions—namely, death—but on the other hand, their intuitive religious conception was that the mental essence of the individual remained. Thus, our minds were conflicted between thinking about the deceased as annihilated and surviving death. Nevertheless, due to one or another of the egocentric mechanisms already discussed, the religious conception of death, researchers claimed, won out in the end.

The questions I want to ask now are whether the experimental results support the theoretical interpretations provided by scientists. Let me be clear: I will not be challenging the developmental, cross-cultural, replicated results that these scientists have offered. On the contrary, I think the results are spot on. What I want to ask is whether the experiments support any of the egocentric theories presented, whether humans (the folk) have these conflicting conceptions of death, and what role, if any, does the fear of death play?

Revisiting the folk psychology of souls

To begin, none of the experimental results demonstrate that I intuitively believe that *I* will survive my own death. Contrariwise, if the results prove that individuals (regardless of age) have *any* intuitive belief about surviving death, it is that individuals intuitively believe that *others* survive death. As I (2011b) previously argued, the current interpretations of these experiments leave an explanatory gap between how I make a non-inferential step from my own personal immortality as posited by the various proposed egocentric psychological mechanisms to the immortality of others as demonstrated by the experimental findings, and such a step must be non-inferential to be intuitive. To date, only one experiment (Pereira et al. 2012) has addressed how individuals view their own death. This experiment, however, asked participants to engage in a conscious, focused, and reflective exercise to imagine themselves after their own death.

Given the findings that we intuitively believe that others can survive death, it is immediately obvious why egocentric explanations, including PCW, will not suffice. The way by which I identify others as the same individual again is not the same as how I identify myself, let's say, after awaking from a dreamless sleep. Put simply, I do not have access to another's mind in any manner similar to which I have access to my own. Instead, I have access to the other's intentions demonstrated by her (previous and present) actions. Thus, in similar fashion, the way by which I would identify myself (even imaginatively) in the afterlife will be different from the way by which I would identify others (also, even imaginatively) in the afterlife. The criteria for both recognition and identification between the two instances are entirely different. Therefore, again, we cannot get from trying to imagine myself surviving death to imagining others surviving death in any intuitive fashion.[12]

With regard to the competing biological and religious conceptions of death, I think there has been a fundamental confusion taking place in these discussions. The researchers appear to have confused the biological conception of death with the scientific/secular conception of death. While the latter posits the annihilation of the individual at the time of death, the former does not. The folk biological conception of death—that is, how we identify *mere* dead bodies—only contains information about how the physical body has met its inevitable death in an irreversible fashion by which all biological functions of that physical body have completely ceased (Speece and Sandor 1984; Barrett and Behne 2005). There is nothing in the folk biological conception of death which, in any way, necessitates understanding the individual whose body that was as annihilated. While this might sound as if I am falling back on some sort of dualism, I am not. All I am saying is simply that there is *nothing* in the biological conception of death that speaks to whether it is possible to survive death or not. The *only* information that conception contains pertains specifically to identifying dead bodies; it says nothing, and suggests nothing, whatsoever about the identity relationship between individuals and their bodies.[13]

The scientific/secular conception of death, on the other hand, states plainly that once all physical functions of the body have ceased that the individual manifested by those physical processes has ceased to exist. It is not in any way clear whether all humans have this scientific/secular conception of death, let alone that they all have it intuitively. Moreover, given the results of the experiments we are considering, it is highly unlikely that "the folk" hold this conception of death at all. The epistemological ability to recognize dead bodies (the folk biological conception) does not necessitate the additional ontological belief mandated by the scientific/secular conception that the individual associated with that body has ceased to exist. If the folk conception of biological death really was the same as the scientific/secular conception of death, then I daresay that the results of those experiments would have been very different, and they certainly would not provide evidence that humans intuitively believe that others survive death. In fact, it would make it all the more difficult to explain how any afterlife beliefs got a foothold in the human mind, let alone one form or another of them being universal across the time and space of human cultures. Afterlife beliefs would have been a nonstarter. There is no immediate conflict, however, between believing that all physical bodies will eventually and inevitably die, never reanimate, and decompose, and that individuals associated with those bodies have survived. It does not have to get any more complicated than that for the folk, and both can be simultaneously intuitive.

With regard to the role that the fear of death plays in afterlife beliefs, Bering (2002) found no relationship between how a participant scored on a Death-Anxiety Scale and their responses about the continuation of certain states after death, but that is not the finding that should evoke fear in TMT theorists. If we are to believe that one of the major (at least) motivational factors for humans to create and adopt afterlife beliefs is because we are terrified of our own demise, then TMT theorists need to explain how and why the youngest children in these studies (4- and 5-year-olds) believed that the characters from the narrative survived death. It has long since been demonstrated in psychology that it is not until ages 7–10 that either children have a fully developed biological conception of death (but as we saw above, that does not really matter), or that they have any self-awareness that they, themselves, or close friends and family will die (Speece and Sandor 1984; Norris-Shortle et al. 1993; Bloom 2004; Cox et al. 2005). This finding, on its own, clearly demonstrates that afterlife beliefs are generated and accepted independently to one's own fear of one's death, and that afterlife beliefs are prior to it. Therefore, TMT and the general assumption held by both philosophers and scientists across the ages that humans have and hold afterlife beliefs because they fear death is unsubstantiated, indeed contradicted, by the evidence.

Reframing the folk psychology of souls around the death we fear

To my mind, there is a better theory that accounts for the findings just discussed. Not only do I believe that the theory I will offer better accounts for the results, but I will also argue that the theory I propose eliminates the current explanatory gap between the findings and the theories employed to explain them and explains the discrepant results produced by minor manipulations in designs employed by different researchers.[14] Moreover, unlike the theories previously discussed, this theory will not require the positing of any novel cognitive mechanism. Again, to be clear: I am not challenging the results of any of these experiments. In what follows, I proffer an alternative empirically testable theory that better accounts for the results already garnered.

I propose an explanation that easily and neatly accounts for all the experimental results found when humans are confronted with the death of another (particularly a loved one, but especially not one's self). This theory employs two workaday psychological mechanisms when thinking about a deceased other, the intentional stance, and offline social reasoning.[15] The natural cognitive outcome of those two mechanisms working together in thinking about deceased others is that they produce, by necessity, a socially embodied representation of that deceased individual in the mind of the living.[16] Socially embodied representations, as I have explained in much greater detail in earlier work (Hodge 2011a), are imaginative images automatically created in our minds when we think about an absent conspecific. The image contains all of the necessary embodied characteristics needed for us to recognize the image as the person being thought about, and for him to continue social interactions not only with us, but also with others, albeit these are still imagined interactions.[17] We use offline social reasoning every day to anticipate, to plan, and to predict future interactions with our conspecifics. The image that we use to think about decedents is created (imagined) in the same way as the ones we use to think about absent conspecifics (i.e., those we think about who are currently outside of our perceptual awareness).[18] Additionally, this means that the deceased individual is and must be imagined *somewhere* doing *something*. This also means that by default, our representations of the deceased in these circumstances are of the deceased in a different location, and this in itself is why humans do not view biological death of the body as the annihilation of the individual. The individual still exists, just not here. Let me provide an example of what this theory would normally look like in practice:

> Sally is confronted with the death of her grandfather with whom she was close. For several years, Sally had lived on the other side of the country from her grandfather. She missed her grandfather and often thought about him. She would remember times he would play with her when she was a child, and imagine how good it would be to see and talk with him face-to-face. If it was a Friday, Sally knew that he would be on the golf course, and she would often imagine him teeing off as he had done the many times she had played golf with him. But now her grandfather was *gone*; he was dead: And Sally is flying home to attend his funeral. During the flight, Sally realizes that she will no longer be able to play golf with her grandfather and a wave of grief overcomes her. She tries to stifle her public sobs by imagining that her grandfather is *looking* down on her *smiling*. She imagines how glad he is to see his family and friends that had died before him, especially his wife, Sally's grandmother, who had

> passed several years before. Sally imagines him holding her grandmother's hand as he is telling Sally that he is okay. And though he misses her too, he will keep a watchful eye on her—or so Sally imagines. Sally silently vows to herself that she will not forget her grandfather, and she is sure that he has heard her vow.

This is an example with which any of us who have lost a loved one, regardless of whether we believe in an afterlife or what kind, can relate.[19] We cannot stop those kinds of thoughts, emotions, and imaginings. They just happen.

Furthermore, this example points out many aspects that can be handled under the theory I am proposing but not by those previously offered. First, when Sally is thinking about/imagining her grandfather, she is not thinking about his dead insensate corpse that they will soon be burying. Sally is thinking about her grandfather the way she always did when they were apart. It is no coincidence that all of our metaphors for death include a change in location (e.g., he has passed, he is gone, he is departed; see also Hodge 2008, 2011a). Second, Sally does not (try to) imagine her grandfather as a disembodied mind. Instead, she uses much the same imaginative representation created through offline social reasoning as she always did. She can tell by the *smile* she imagines on his face, from the words she imagines him *telling* her, and by the grasp he has on her grandmother's *hand* that he is okay, and he intends to watch over her. In Sally's mind, her grandfather has transitioned from this life to the afterlife. But even though Sally's grandfather is gone, she continues to think about him and imagine him in ways that continue his social bonds and interactions not only between him and Sally, but also between him and others both living and deceased (see, Boyer 2001; Bering 2006, 2010; Dennett 2006 for similar examples of these phenomena).

Moreover, as Sally's mental representation of her grandfather fades, as it is bound to do over time, Sally *fears losing him forever* (or at least until the end of her life, anyway). In anticipation, and fear, of his image fading from her mind, Sally vows to remember him. This is one of the most common fears expressed by those who have lost a loved one (Norris-Shortle et al. 1993; Wilkin and Powell 1996; Northern Arizona Healthcare 2010). I have previously called the loss of memory of a deceased loved one a social death (Hodge 2012, following, Palgi and Abramovitch 1984, also called "death after death," see Schmidt 2000), and it is this fear, called athazagoraphobia, which humans fear more than death itself, their own included.[20] By forgetting those we loved after their death, we have annihilated them from our minds and thus existence—or so it seems.

The trauma of losing a loved one to death is one of the most emotionally tumultuous experiences humans can face. The strong, visceral impact that these representations have on us presents as real, and thus we treat them as real and as true (Hodge 2011a). Moreover, religious beliefs such as afterlife beliefs can aid us in returning to emotional homeostasis during traumatic events (Schjødt 2007).[21] Perhaps that is why even atheists have been caught-out in these experiments affirming that some states continue (Bering 2002, 289, 2006, 3, 2010, 117–118). At a minimum, it explains why even atheists can and do imagine scenarios similar to the above when faced with the death of a loved one.[22] Contra Richard Dawkins (2006), such representations and beliefs are not merely wishful thinking due to the fear of death. They are our mind's natural and intuitive manner of dealing with such great loss.

But let me turn away from my example now and return to other ways in which the theory I offer here provides better coverage for the results. One puzzling result that Bering and colleagues reported was that while children were significantly less likely to claim that the deceased subject of the narrative, in these cases a puppet baby mouse, would eat now that he was dead, they found it significantly more likely that the baby mouse might be hungry (Bering and Bjorklund 2004; Bering et al. 2005). Clearly, this was a finding that did not sit well with mind-body dualism. Additionally, another problem for Bering and colleagues to fit the findings with a mind-body dualism was that perceptual abilities were not regularly said to continue for the deceased.[23] Under the theory I am proposing, however, both of these oddities are easily expected and explained by the intentional stance. Intentional versus non-intentional states do not break in the same way as one would expect mental versus physical to do, especially in comparison to Cartesian substance dualism orthodoxy.[24] While "being hungry" is not (purely) a mental state, it is an intentional state. And while Bering and colleagues found that perceptual states of the decedent ceased for smells and tastes with nonsocial significance, Bek and Lock (2011) and Huang et al. (2013) found that both visual and auditory states were more likely to be said to continue over olfactory, tactile or gustatory states, especially if something of social significance was what was being seen or heard, thus causing the imaginer to take the intentional stance toward those perceptual abilities.[25] More recent experiments by Lane et al. (forthcoming) provide additional confirmation of how manipulating the social significance of these questions can have a significant effect on whether any states, including so-called mental states, are believed to continue after death.

And finally, the social embodiment theory fits well with the common "folk" representations of decedents present in anthropological findings, mythologies, literature, and movies. This, we learned from Flew, is the ABW. We continue to think about and represent the deceased in an embodied manner, so that they can continue to interact with us, the living, and their fellow deceased. It is okay if their representations range from the clear vision to the fuzzy apparition—that is the nature of both memory and imagination. In contradistinction, however, PCW apparitions have never appeared to anyone other than philosophers and a few psychologists. As Jaegwon Kim (2001) once described it, if the deceased were disembodied Cartesian minds, they would be very lonely souls.

Philosophical implications of the folk psychology of souls

I began this chapter discussing how the assumption that belief in an afterlife was at most caused and at least motivated by the fear of my own death has gone largely unchallenged in both philosophy and science. The challenge has occurred, and the assumption has lost. While it is still an interesting question as to whether we should fear our own death, philosophers should not expect people's afterlife beliefs to be influenced or changed based on such analysis.

The Epicurean decree that (our own) death is nothing to us may be true, but the death of our loved ones certainly *is* something to us. It is, and should be, a traumatic, emotionally wrought, experience. The afterlife beliefs we hold (intuitively or otherwise) are there to ameliorate our grief and the fear the loss of a loved one creates in our lives. They are not created or designed to buffer our anxiety toward our inevitable annihilation. In fact, the evidence suggests that humans do not view death as an annihilation, but rather a change in location. Even worse for the Epicureans is that there is now cross-cultural evidence that children do not view their prelife as nonexistence, but rather as socially engaged beings (Emmons and Kelemen 2014). Nevertheless, it is incredibly handy that by the time we become mature enough to realize that we too will die, our minds have already created a place for us to inhabit alongside all our deceased loved ones.

There is no doubt that the prospect of our own death and prospective (even probable) annihilation, when we attempt to dwell on it with any intensity, can be terrifying. Upon reflective scrutiny, our intuitive afterlife beliefs will likely not hold up too well, so there is still a time for the consolation of philosophy. Wanting to approach our impending death with a calm that comes with wisdom and having lived the good life are certainly noble pursuits. So too is hoping (even against hope) that we will see and be with our loved ones again.

Psychologists are, and have been, treading into topics long considered to be the domain of philosophy—whether it be morality, the nature of the mind, personal identity, or belief in the supernatural. Philosophers should attempt to keep themselves well informed on the latest empirical findings in their fields of interest. But, more important than that, philosophers should make every attempt to engage in the theoretical debates scientists are having about how to explain the empirical findings. This I see as vitally important to help keep up the pace of science's progressive march. Why? Because scientists are trying to avoid, yet sometimes heading straight into, many of the philosophical dead-ends that have long ago been discovered by philosophers.

In the recently emerging field of cognitive science of religion, to which this book is certainly a contribution, philosophers of mind and religion can help scientists avoid logical errors that have be discovered, fill explanatory gaps that have been uncovered, and keep them from committing philosophical howlers (Dennett 2007). As entertaining as these latter might be, they certainly are of little help.

Help Wanted: Philosophers Apply Within.

Acknowledgments

I would like to thank the many who gave helpful feedback and comments on earlier drafts of this chapter: Graham Macdonald, Claire White, Louise Tardiff, Justin Lane, Don Braxton, Rajiya Jones, Kelly Clark, Claire McMahon, and an anonymous reviewer. I would also like to thank Helen De Cruz and Ryan Nichols for both their feedback and their invitation to be a part of this project.

Notes

1 The phrase "survive death" is very odd. I have heard that phrase, and those like it for most of my life, but it was not until a few years ago that I suddenly realized how self-contradictory that phrase is, as Flew (1956, 1967/2006) I later found had long since pointed out. And this was already after investigating views of death and the afterlife for many years! To some, it might be definitive proof that I am thick, but to others, it points out how seductive and intuitive the idea that one can survive death *really* is.

2 For overviews of the Epicurean position on death and its nothingness to us, see Kaufman (2004), Martin and Barresi (2006), Rosenbaum (2004), Sorabji (2006).

3 Flew varied his terminology for these three positions: in some works, he referred to these views as doctrines instead of ways, and in others, he referred to the PCW as "the immortal soul doctrine" and the ABW as "the shadow man doctrine." The presentation of these views, however, varied little despite the name changes. In his final work on this topic in 1998, Flew had settled back into the terminology I use here.

4 Albeit this is a healed and perfected body.

5 Hick (1978, 2001), Murphy (2006), and van Inwagen (1995) being a notable exceptions.

6 Additional psychological/cognitive explanations are provided by Harris and Giménez (2005), Astuti and Harris (2008), Bek and Lock (2011), Richert and Smith (2012), and Pereira et al. (2012). I do not address them here because I believe they are sufficiently similar enough to the theories I will discuss that my criticisms ultimately apply to them as well.

7 This phrase to describe folk afterlife beliefs was introduced by Bering (2006).

8 Contrary to a common reading, the work of Astuti, Gimenez, and Harris does not challenge the finding that belief in an afterlife is intuitive: see Hodge (2012, *n.* 4).

9 With regard to perceptual abilities, Bek and Lock (2011) and Huang et al. (2013) found that continuation responses for perceptual abilities would increase significantly with a change in context, specifically a social context such as "Now that **S** is dead, can **S** still see person *P*?" Such contextual results for these manipulations were predicted by Hodge (2008, 2012) and will be discussed again briefly below. Another interesting result in these studies was that continuation responses for psychobiological abilities remained high. I will address this in more detail below.

10 This is not to say that Theory of Mind is inherently dualistic as a theory, but merely it is this theory that psychologists used to make the division between mental states and physical states. Applications of Theory of Mind in discussions of empirical findings, however, often assume/suppose dualism. That discussion is beyond the scope of this chapter.

11 I have always found it interesting, as well as a little belittling, that scientists are claiming that humans, including children, intuitively hold and believe a position that took centuries of careful and deliberative philosophical investigation into the nature of personal identity to develop. Cartesian substance dualism, at least until a century ago, was a position believed to be reserved for the well-studied academic. Now, scientists would have us believe that the position carefully reflected upon and argued for by some of humankind's greatest minds was simply the exegesis of naïve and intuitive notions which even children grasp!

12 Of course, *if there is* an afterlife, it is possible that it could be a holding cell for disembodied minds who both identify and interact purely through some sort of

mental telepathy, but is that really what any of these researchers are suggesting that these participants are intuitively representing?

13 According to Barrett and Behne (2005), the mechanisms under girding biological conception of death solved an important evolutionary problem: namely, distinguishing between intentional agents and non-agents. This is because the potential costs of failing to discriminate between the two are high: it could result in the loss of one's life, produce unnecessary vigilance, or a lost opportunity for food.

14 See both Hodge (2011b) and (2012) for additional discussion of the issues raised here.

15 I also think that this proposed theory is more parsimonious, even though it requires more moving parts, given that it does not require any beyond necessity.

16 In the interest of full disclosure, I am not the first nor the only researcher into afterlife beliefs that have proposed a theoretical role for either offline social reasoning or the intentional stance. For instance, Bering (2006) suggested the importance of offline social reasoning in thinking about deceased others—though he did not suggest it was a causal mechanism. Additionally, Daniel Dennett (2006) suggested how we use the intentional stance to think about the deceased. Theorizing that the deceased are represented in a socially embodied way, however, is my own (as are any faults with the proposed combination of these elements).

17 My theory of social embodiment has received additional support when thinking about supernatural agents such as gods, even though it has not yet been directly tested on afterlife representations. The experiment (De Cruz 2013) demonstrated that participants were more likely to represent gods as socially embodied over disembodied minds and broad anthropomorphism.

18 Although there is more to the story when thinking about fictional characters, certainly the intentional stance, offline social reasoning, and social embodiment all play a role.

19 Assuming, of course, that we are all neurotypical. Also see note 16 above.

20 This is also why cults of the dead play such a prominent role in world religions.

21 I wish to be perfectly clear, however, that I am in no way suggesting that either the ability to take the intentional stance or offline social reasoning evolved to aid us in dealing with the deaths of our loved ones. I think it is just a happy accident that they do.

22 Both Hodge (2010) and Bering (2010) provide specific anecdotal evidence of this phenomenon.

23 Bering (2002, 2006) explained this by stating that people reasoned that such perceptual abilities were dependent upon their physical mechanisms (i.e., eyes and ears, etc.). But if that is the case, why is it so intuitively natural to hear someone say that their deceased loved one is watching over them? How is that possible under Bering's reasoning?

24 Under Cartesian substance dualism, perceptual states are counted among mental states, and emotional states are counted as physical/bodily states. Yet, in some interpretations of these findings, such as Bering (2006), perceptual states are asserted to be physical/bodily states and emotional states are asserted to be mental states. The findings, however, show perceptual states (especially when introduced in the narrative on their own—i.e., no social context) to have a much lower continuity response ratio than do emotional states. This is the opposite of what Cartesian substance dualism would predict. To circumvent this problem (if it is even recognized as a problem), researchers have used Theory of Mind to divvy up the states between mental and physical. For a detailed discussion of the difference between Theory of Mind which supports the mental state/physical state divide invoked by most of the theories provided here and the Intentional Stance which treats mental states as useful fictions, see Griffin and Baron-Cohen (2002). As much as I would like to discuss this more here, it is truly beyond the scope of this chapter.

25 This is not to say that smell, touch, and taste cannot play important roles in some social interaction between humans, but they are not our dominant sensory modes of recognition of others.

Bibliography

Astuti, R. 2011. Death, ancestors, and the living dead: Learning without teaching in Madagascar. In V. Talwar, P. L. Harris and M. Schleifer, eds, *Children and death: From biological to religious conceptions*, 1–18. Cambridge: Cambridge University Press.

Astuti, R. and P. L. Harris. 2008. Understanding mortality and the life of the ancestors in rural Madagascar. *Cognitive Science* 32: 713–740.

Barrett, H. C. and T. Behne. 2005. Children's understanding of death as the cessation of agency: A test using sleep versus death. *Cognition* 96: 93–108.

Becker, E. 1973. *The denial of death*. New York: Free Press.

Bek, J. and S. Lock. 2011. Afterlife beliefs: Category specificity and sensitivity to biological priming. *Religion, Brain & Behavior* 1: 5–17.

Benatar, D., ed. 2004. *Life, death, & meaning: Key philosophical readings on the big questions*. Oxford: Rowman & Littlefield Publishers, Inc.

Bering, J. 2002. Intuitive conceptions of dead agents' minds: The natural foundations of afterlife beliefs as phenomenological boundary. *Journal of Cognition and Culture* 2: 263–308.

Bering, J. 2010. *The god instinct: The psychology of souls, destiny and the meaning of life*. London: Nicholas Brealey Publishing.

Bering, J. and D. F. Bjorklund. 2004. The natural emergence of reasoning about the afterlife as a developmental regularity. *Developmental Psychology* 40: 217–233.

Bering, J., C. H. Blasi, and D. F. Bjorklund. 2005. The development of "afterlife" beliefs in religiously and secularly schooled children. *British Journal of Developmental Psychology* 23: 587–607.
Bering, J. M. 2006. The folk psychology of souls. *Behavioral and Brain Sciences* 29: 1–46.
Bloom, P. 2004. *Descartes' baby: How child development explains what makes us human.* London: Arrow Books.
Bloom, P. 2007. Religion is natural. *Developmental Science* 10: 147–151.
Boyer, P. 2001. *Religion explained: The evolutionary origins of religious thought.* New York: Basic Books.
Cox, M., E. Garrett and J. A. Graham. 2005. Death in Disney films: Implications for children's understanding of death. *Omega* 50: 267–280.
Dawkins, R. 2006. *The god delusion.* London: Transworld Publishers.
De Cruz, H. 2013. Religious concepts as structured imagination. *International Journal for the Psychology of Religion* 23: 63–74.
Dennett, D. 2007. Philosophy as naive anthropology: A comment on Bennett and Hacker. In M. Bennett, D. Dennett, P. Hacker, and J. Searle, eds, *Neuroscience & philosophy: Brain, mind, & language*, 74–95. New York: Columbia University Press.
Dennett, D. C. 2006. *Breaking the spell: Religion as a natural phenomenon.* New York: Viking Penguin.
Descartes, R. 1641/1993. *Meditations on first philosophy.* Indianapolis, IN: Hackett Publishing Company, Inc.
Emmons, N. A. and D. Kelemen. 2014. The development of children's prelife reasoning: Evidence from two cultures. *Child Development* 85: 1617–1633.
Flew, A. 1953. *A new approach to psychical research.* London: Watts.
Flew, A. 1956. Can a man witness his own funeral? *Hibbert Journal* 54: 242–250.
Flew, A. 1967/2006. Immortality. In D. M. Borchert, ed., *Encyclopedia of philosophy* (2nd ed.): 602–616. Farmington Hills, MI: Thomson Gale.
Flew, A. 1998. Could we survive our own deaths? *Internet Infidels* [Online]. Available: http://www.infidels.org/library/modern/antony_flew/survival.html [Accessed July 9, 2013].
Goldenberg, J. L. 2005. The body stripped down: An existential account of the threat posed by the physical body. *Current Directions in Psychological Science* 14: 224–228.
Greenberg, J., T. Pyszczynski, S. Solomon, A. Rosenblatt, M. Veeder, S. Kirkland, and D. Lyon. 1990. Evidence for terror management theory II: The effects of mortality salience on reactions to those who threaten or bolster the cultural worldview. *Journal of Personality and Social Psychology* 58: 308.
Griffin, R. and S. Baron-Cohen. 2002. The intentional stance: Developmental and neurocognitive perspectives. In A. Brook and D. Ross, eds, *Daniel Dennett: Contemporary philosophy in focus,* 83–116. Cambridge: Cambridge University Press.
Harris, P. L. and M. Giménez. 2005. Children's acceptance of conflicting testimony: The case of death. *Journal of Cognition and Culture* 5: 143–164.

Healthcare, N. A. 2010. *10 stages of grief* [Online]. Flagstaff: Northern Arizona Healthcare. Available: http://www.nahealth.com/OurServices/NorthernArizonaHospice/EmotionalandSpiritualSupport/10_Stages_of_grief [Accessed May 18, 2010].

Hick, J. 1978. Present and future life. *The Harvard Theological Review* 71: 1–15.

Hick, J. 2001. Resurrecton of the person. In M. Peterson, W. Hasker, B. Reichenbach, and D. Basinger, eds, *Philosophy of religion: Selected readings* (2nd ed.): 478–488. New York: Oxford University Press.

Hodge, K. M. 2008. Descartes' mistake: How afterlife beliefs challenge the assumption that humans are intuitive Cartesian substance dualists. *Journal of Cognition and Culture* 8: 387–415.

Hodge, K. M. 2010. *Cognitive foundations of afterlife beliefs*. Ph.D. Thesis, Queen's University Belfast.

Hodge, K. M. 2011a. On imagining the afterlife. *Journal of Cognition and Culture* 11: 367–389.

Hodge, K. M. 2011b. Why immortality alone will not get me to the afterlife. *Philosophical Psychology* 24: 395–410.

Hodge, K. M. 2012. Context sensitivity and the folk psychology of souls: Why Bering *et. al.* got the findings they did. In D. Evers, M. Fuller, A. Jackelén, and T. Smedes, eds, *Is religion natural*? 49–63. New York: T & T Clark International.

Huang, J., L. Cheng, and J. Zhu. 2013. Intuitive conception of dead persons' mentality: A cross-cultural replication and more. *International Journal for the Psychology of Religion* 23: 29–41.

Kaufman, F. 2004. Pre-vital and post-mortem non-existence. In D. Benatar, ed., *Life, death, & meaning: Key philosophical readings on the big questions*, 243–266. Oxford: Rowman & Littlefield Publishers, Inc.

Kim, J. 2001. Lonely souls: Causality and substance dualism. In K. Corcoran, ed., *Soul, body, and survival: Essays on the metaphysics of human persons*, 65–77. Ithaca, NY: Cornell University Press.

Kuhlmeier, V. A., P. Bloom, and K. Wynn. 2004a. Do 5-month-old infants see humans as material objects? *Cognition* 94: 95–103.

Kuhlmeier, V. A., K. Wynn, and P. Bloom. 2004b. People V. objects: A reply to Rakison and Cicchino. *Cognition* 94: 109–112.

Lane, J. D., L. Zhu, E. M. Evans, and H. M. Wellman. Forthcoming. Developing concepts of the mind, body, and afterlife: Exploring the roles of narrative context and culture. *Journal of Cognition and Culture.*

Martin, R. and J. Barresi. 2006. *The rise and fall of soul and self: An intellectual history of personal identity*. New York: Columbia University Press.

Murphy, N. 2006. *Bodies and souls, or spirited bodies*? Cambridge: Cambridge University Press.

Nichols, S. 2007. Imagination and immortality: Thinking of me. *Synthese* 159: 215–233.

Norris-Shortle, C., P. A. Young, and M. A. Williams. 1993. Understanding death and grief for children three and younger. *Social Work* 38: 736–741.

Palgi, P. and H. Abramovitch. 1984. Death: A cross-cultural perspective. *Annual Review of Anthropology* 13: 385–417.

Pereira, V., L. Faísca, and R. D. Sá-Saraiva. 2012. Immortality of the soul as an intuitive idea: Towards a psychological explanation of the origins of afterlife beliefs. *Journal of Cognition and Culture* 12: 101–127.

Pyszczynski, T., J. Greenberg, and S. Solomon. 1999. A dual-process model of defense against conscious and unconscious death-related thoughts: An extension of terror management theory. *Psychological Review* 106: 835–845.

Richert, R. A. and E. I. Smith. 2010. The role of religious concepts in the evolution of human cognition. In U. Frey, ed., *The nature of god–evolution and religion*, 93–110. Antwerp: Tectum.

Richert, R. A. and E. I. Smith. 2012. The essence of soul concepts: How soul concepts influence ethical reasoning across religious affiliation. *Religion, Brain, & Behavior* 2: 161–176.

Rosenbaum, S. E. 2004. How to be dead and not care: A defense of Epicurus. In D. Benatar, ed., *Life, death, & meaning: Key philosophical readings on the big questions*, 175–189. Oxford: Rowman & Littlefield Publishers, Inc.

Schjødt, U. 2007. Homeostasis and religious behavior. *Journal of Cognition and Culture* 7: 313–340.

Schmidt, B. 2000. Memory as immortality. *Near Eastern Archaeology* 63: 236–239.

Solomon, S., J. Greenberg, and T. Pyszczynski. 1997. Return of the living dead. *Psychological Inquiry* 8: 59–71.

Sorabji, R. 2006. *Self: Ancient and modern insights about individuality, life, and death.* Oxford: Clarendon Press.

Speece, M. W. and B. B. Sandor. 1984. Children's understanding of death: A review of three components of a death concept. *Child Development* 55: 1671–1686.

Vail Iii, K. E., Z. K. Rothschild, D. R. Weise, S. Solomon, T. Pyszczynski, and J. Greenberg. 2010. A terror management analysis of the psychological function of religion. *Personality and Social Psychology Review* 14: 84–94.

Vallacher, R. R. 1997. Grave matters. *Psychological Inquiry* 8: 50–54.

van Inwagen, P. 1995. Dualism and materialism. *Faith and Philosophy* 12: 475–488.

Wilkin, C. S. and J. Powell. 1996. *Learning to live through loss: Helping children understand death* [Online]. Manhattan, KS: National Network for Childcare/Kansas State University Cooperative Extension Service. Available: http://www.nncc.org/Guidance/understand.death.html [Accessed May 18, 2010].

Index

www.ingramcontent.com/pod-product-compliance
Ingram Content Group UK Ltd.
Pitfield, Milton Keynes, MK11 3LW, UK
UKHW020141290926
15471UKWH00003B/621

9 781350 041561